About the author

Steve Martin is a classically trained businessman who spent his early career in the confectionery industry. There, he was given a conventional business training, which included a grounding in financial controls, checks and balances, ethical business practice, budgetary control, sales and marketing, and business leadership.

He has worked in the sign industry for over 15 years. On first entering the industry, he was shocked at the low level of management training, and the fact that a large proportion of managers in the industry have virtually no formal training in the technical aspects of management.

Many firms were managed without reference to key ratios, and few managers understood the concept of "strategy" (though many used the word, and applied it incorrectly). Adherence to the law, health and safety, quality, and ethical good business practice was restricted to a small number of companies.

Shortly after the millennium, he founded Xmo Strata, and within three years, this company had become the independent UK leader in sign installation and maintenance. The company has the most highly trained sign installation engineers in the UK, and the most comprehensive health and safety policies – some of them borrowed directly from the aerospace industry.

He expects "heavy flak" upon publication of this book, but is so deeply angered by the scale of deception within the sign industry that his attitude is uncompromising: "if there's going to be flak, bring it on," he says. "No company, in any field, anywhere in the world, is perfect. We all make mistakes from time to time, because we're all human. But the level of institutionalised bad practice, incompetence, and downright fraud in the sign industry is not a question of occasional mistakes and misjudgements, it is frequently a question of organised deception - and it has to be stamped out.

"There *are* good companies, and good individuals in the industry - many of them – but there is no point in saying that the failures are down to a few bad apples, because they are not. This is happening on an alarming scale. Some in the industry will probably adopt a denial mode about the book – I suspect that the customers who read it will be somewhat less sanguine, however."

Although Xmo Strata is a UK-wide firm, Steve Martin is a born and bred Kentishman.

*Declaration of interest: the author owns and runs the largest independent sign installation and maintenance company in the UK.

Safety, quality, tricks, and lies

Dirty tricks in the British sign industry – and 100 questions your sign company doesn't want you to ask!

By Steve Martin

Published 2007 by arima publishing

www.arimapublishing.com

ISBN 978-1-84549-172-7

© Steve Martin 2007

All rights reserved

This book is copyright. Subject to statutory exception and to provisions of relevant collective licensing agreements, no part of this publication may be reproduced, stored in a retrieval system, or transmitted in any form or by any means, without the prior written permission of the author.

Printed and bound in the United Kingdom

Typeset in Palatino Linotype 11/14

This book is sold subject to the conditions that it shall not, by way of trade or otherwise, be lent, re-sold, hired out, or otherwise circulated without the publisher's prior consent in any form of binding or cover other than that in which it is published and without a similar condition including this condition being imposed on the subsequent purchaser.

arima publishing
ASK House, Northgate Avenue
Bury St Edmunds, Suffolk IP32 6BB
t: (+44) 01284 700321

www.arimapublishing.com

Cover: Eclipse Design www.eclipse-design.co.uk

Acknowledgements

Many people have helped in the research, writing, and editing of this book. Some of those in the sign industry have done so on conditions of strict anonymity, which I will always respect. I think it is a sad reflection on the nature of the industry that those who have provided information on activities which are illegal, immoral, or unethical, do not feel in any way reassured by the legal protection which is now, supposedly, afforded to 'whistle blowers'. One or two of them have been pressured by their companies into practising some of the less ethical activities described in the book and, therefore, feel that they share culpability. You know who you are; it is never easy to blow the whistle, even anonymously, and I hope you feel better for having done so. This is the only way I can thank you, inadequate though it is. Any mean-spirited company bosses who scrutinise the copy for clues as to your identity will, I hope, be frustrated.

Before I thank any named individuals, I should make it very clear that the tone and content of the book are entirely mine, as are any mistakes. If you write a book with a 'robust' tone, you must be prepared to take the consequences, for the simple reason that if you can't take it, you have no business dishing it out!

Having said that, it would be unwise to undertake a book like this without having sound legal advice, and the manuscript has been scrutinised by *two* barristers.

Two of the senior managers in my own company, Michael Mott and Kate Spencer-May, read the copy and provided excellent criticism, which helped improve the book considerably. That was over and above the call of duty, guys.

The remainder of the companies, organisations and web sites mentioned in the book are there for the assistance and information of the reader; they have not endorsed the book, or supported it in any way, and their presence in the copy does not imply anything other than the fact that I decided to include them, because I thought that doing so might assist the readers. For the most part, they had no prior knowledge that they would be included.

Of equal importance, I do not in any way endorse *them*; I merely offer the contact.

Contents

FOREWORD   9

PREFACE   13

CHAPTER 1
    Credentials   15

CHAPTER 2
    Financials and Sub-Contracting   35

CHAPTER 3
    The Contract   43

CHAPTER 4
    Project Management and Contract Implementation   49

CHAPTER 5
    Health & Safety   57

CHAPTER 6
    Dirty Tricks   79

CHAPTER 7
    Be a good customer   95

CONCLUSION
    Don't have nightmares   107

APPENDICES
    1. A quick checklist and matrix   109
    2. Dirty Tricks checklist   115
    3. Useful website addresses   117

Foreword

This book is going to upset a lot of people, and whilst that is not my purpose in writing it, nor will it deter me. The thrust of the book is that you should exercise great care in doing business with sign companies. All sign companies are not rogues, however; there are many good companies. I sincerely hope that the book will help customers to single them out, to the commercial disadvantage of the remainder.

For those who buy a lot of signage, a book like this might be a useful aide-memoir. They are probably aware of the endemic problems in some parts of the industry, alert to the risks, and adept at handling them. Even some of the experts, however, might find one or two twists in these pages that are new to them. I certainly found a few, when researching the book.

For those who haven't bought signs before, however, the book should act as a very large warning. Do not go into the business of buying signage lightly; be well prepared. I repeat, it is not true to say that every sign company is managed by sharks and staffed by spivs – there are some very good firms, offering an excellent service – but the industry, as a whole, does suffer from a culture of low business and technical skills, and price-led selling.

Not all the 100 questions you see here will necessary apply to your situation; you will have to sift out those that are relevant. No one is suggesting that it is practical, or sensible, to subject a sign company to the full interrogation that would be required if you asked all 100 questions given here – life's too short!

Conversely, an awful lot of these questions are not exclusively applicable to the purchase of signs. If you have taken the trouble to buy a book like this, I assume you are a manager in one of the companies that regularly spend money on signage. But you will, doubtless, buy other goods and services, as well, and I hope the book serves a wider purpose.

Big brand companies tend to be involved in a large signage project whenever there is a new brand or corporate identity launch, new product launches, or big promotions; but even if you run a corner shop, and buy one sign every 10 years, the principles will still apply.

The entry-threshold to the sign industry is low, and consequently, it is hugely over-populated. Of the (roughly estimated) 3,000 or so small UK sign companies, many last for a decade or less, and may be run by people who come from an entirely different sector of commerce (running a cab firm, for instance).

The larger companies that are tempted to transgress do so because they are undercut by low-skills, low-cost-base, high-risk smaller firms, and because, in some cases, the customers are ill-equipped to differentiate between competing claims, and identify those which are hollow.

At the more benign end of the scale, some companies will do nothing more than overstate their capabilities, resources, and experience, and many would argue that this is mere 'salesmanship', a question of putting the best foot forward, and that almost every company, in any sphere, does the same. But the truth is that in the sign industry, many of the more sinister practices described here are commonplace.

FOREWORD

Some readers may worry that a book like this could have the reverse effect of the intended one – in that it might become a how-to-do-it guide for those who want to cut corners, or behave unethically.

I accept that any attempt to 'blow the whistle' on shady practices, in any industry, carries with it the risk that morally dubious companies and individuals will scrutinise it, with a view to finding new ways to fleece their customers. Where possible, in the chapters that follow, I have outlined the problems in order to expose the issue, rather than provide technical details on how it can be done.

More generally, however, my defence is the one traditionally used by journalists – that disreputable business practices thrive in an environment of secrecy and cover-up. By providing good, clear, honest information for companies, by throwing light on the problem areas, we hope to contribute towards raising the level of service across the board.

Finally, a word to the 'nosy neighbours', the sign industry executives, working for firms that compete with my own, who have bought the book to see "what they're up against". There is no intended malice in this book, but I am rather weary of attending sign industry events where the assumption is made that because I am in the industry, I know the score, and everyone does the same, don't they mate, wink, wink.

You will not want to hear this, but I hope that this book makes at least a small contribution to forcing up standards, raising the entry threshold to the industry and (I'm afraid) hastening the demise of those firms who are so mired in bad practice that they are unable to plan their way to a more ethical style of business.

No one, and no company, is perfect, but some try harder than others to achieve high standards. My aim, with the book, is to help customers spot those that do make a genuine effort to meet the highest standards.

Steve Martin
Maidstone
June, 2006.

Preface

It is assumed that the reader is a customer, or potential customer, and the questions which are posed in the book are biased towards those that are most likely to interest a customer.

For example, sign manufacturing operations are full of health and safety hazards; yet few of the questions address this area, because health and safety infringements which take place on the sign company's own premises are unlikely to directly affect the customer. On the other hand, with the current state of both civil and criminal law, infringements which take place on the customer's premises may well place the customer in legal peril.

There is one excellent piece of advice, however: go and see the factory. There are some "virtual" sign companies – they have no manufacturing capability, but in some cases, few of their customers know that. They source signs from where they can in the industry, put a mark-up on, and sell them on to you – without telling you that someone else did the work. If you are using someone as an "agent" to source signs for you, that's fine, and perfectly legitimate – so long as you *know* that this is their role, and are not deceived into thinking that they are making the signs themselves !

A major British household name used a foreign sign company for over 20 years without realising that the company had no manufacturing capability. Foreign manufacturers may also have different working practices, standards, and conditions, which (depending on your own corporate policies) you may need to know about.

Chapter One
Credentials

You can tell a lot about a sign company from its credentials, but they can be misleading. Some shady companies will try to undermine the significance of the certification you are asking about, and some will tell outright lies in order to secure your business. This is why you should always seek hard evidence to back-up credential claims, before you let a sign company loose on your site.

Under-qualified sign companies will airily claim that their crews have a great deal of "experience", but there is no certificate for experience - and anyway, experience of doing something incorrectly is not very helpful. If a firm has the right experience, and its people can genuinely do the job, there is no reason for them not to have the formal paperwork to prove it.

A traffic reporter may spend many years flying daily in the right hand front seat of a light aircraft. He will have plenty of experience of light aircraft, and if push came to shove, he might even be able to make some kind of a fist of an emergency landing (if the pilot had died of a sudden heart attack, say) - but that doesn't mean he's a suitable person to fly you around.

The usual excuse for not having particular credentials is that they are "not necessary", that having employees attend the relevant training courses would "drive up costs", and that these would, ultimately, be passed on to the customer.

My answer to that is that the cost of return visits to do the job properly, because it wasn't done properly in the first place, will be far more expensive than the cost of a few training courses, and that in today's "compensation culture", I wouldn't want anyone working on a site for which I was responsible unless they were clearly qualified to do so.

The objective of this chapter is to help you sift those with genuine credentials, from those with hollow claims.

1. **Have all your sign installation teams completed the PASMA (Prefabricated Access Suppliers' and Manufacturers' Association) Standard Course?**

Completion of The PASMA Standard Course (as a basic minimum) should be required of *anyone* and *everyone* using mobile scaffold towers. Of course, if such towers aren't being used on the project, it isn't necessary – but do ask yourself whether you think they *should* be!

If someone breaches safety and working at height regulations, and there is an industrial accident as a consequence, *your* company could, potentially, be sued by the injured employee, or prosecuted by the HSE, even if you did not employ them directly. Even you, personally, may be put in legal peril.

No-win, no-fee lawyers will go after the richest "culpable" company, or (nowadays) individual, and since many sign companies are very small, they might swerve around them, and sue the customer (or its directors)!

You, as the site owner, must provide a safe place of work – and it is also your responsibility to ensure that competent contractors are used.

Your sign company *may* say that it is sufficient for *one* person, in each crew, to have completed The PASMA Standard Course, but it is up to you to decide whether you regard this as acceptable. It is far better, of course, that each crew member has completed it, and understands the best practice requirements. You can find out more about PASMA at www.pasma.co.uk.

2. **Have your managers completed the IOSH (Institution of Occupational Safety and Health) Managing Safely course?**

Even if it is a small company, it is reasonable to expect at least *one* manager to have completed this course. If no one in the company has such training, there is really no excuse. In a larger company, there should be a number of certificate holders amongst the management team. If, in addition, they have completed *other* IOSH courses, that's all well and good. IOSH is Europe's leading body for health and safety professionals; completion of this course puts an emphasis for safety on managers, as well as site workers. You can find out more about IOSH at www.iosh.co.uk.

3. **Have your installers all completed the IPAF (International Powered Access Federation) Mobile Elevated Work Platform, Safety Harness, and (where appropriate) the Self Propelled Boom category of the MEWP course?**

Ideally, the Mobile Elevated Work Platform and Safety Harness courses should be completed by *all* workers using scissor lifts and powered access equipment. It is essential that workers have a stable platform when working at height, and that they know how to use it.

Note that anyone who uses truck-mounted cherry pickers for their work should have completed the Self Propelled Boom category of the MEWP course.

Falling from height is one of the most common *serious* industrial accidents in the UK, and, surprisingly to some people, falls from less than two metres are amongst the most dangerous, because the victim is more likely to suffer head injuries. You can find out more about IPAF at www.ipaf.org.

4. **Is the company accredited to ISO (International Standards Organisation) 9000?**

Somebody once said that the bitterness of poor quality remains, long after the sweetness of low price is forgotten. A sign represents your brand values, and once your sign is up, these values are there for all to see. ISO 9000 is a quality certification which will help to ensure that your brand values are displayed to the standards you would expect.

CHAPTER ONE - CREDENTIALS

ISO 9000 has been around for many years – so there is no reason for a sign company not to have it. If it doesn't, you are right to draw negative conclusions.

Many companies will try to dismiss quality certification like this as being unnecessary, but it is a proven, auditable, check, on the sign company's quality procedures. It means that there is quality traceability on the signs they make, and it should give you some peace of mind. You can find out more about ISO 9000 at www.iso.org.

5. **If the work involves petrol forecourts, do the installations comply with the 2nd edition of** *Guidance for the Design, Construction, Modification, Maintenance and Decommissioning of Filling Stations* **(sometimes known as 'The Blue Book'), published in 2005?**

The answer here must be 'yes', the company is fully compliant. The Blue Book is 'The Bible' for guidance on construction and work on petrol forecourts. Even a pretty shady company should be aware of this. You can find out more about the Blue Book at www.apea.org.uk.

6. **If the work involves petrol forecourts, have all your crews completed the UK PIA Forecourt Contractor Safety Passport Scheme?**

This is the UK Petroleum Industry Association qualification – sometimes referred to as the "safety passport" – for those working on petrol forecourts. Work on petrol forecourts presents some particular safety issues, and the training is designed to address these. In my view, it would be highly irresponsible to employ contractors of any sort to work on a forecourt without this qualification. You can find out more about UK PIA at www.ukpia.com.

7. **If electrical work is carried out on petrol forecourts, or hazardous areas, have all operatives completed the Comp'Ex 07 and 08 courses?**

Comp'Ex is the leading national safety training scheme for electricians working in the potentially hazardous or explosive atmospheres. Potentially explosive atmospheres occur in both onshore, and offshore petrochemical plants, and refining plants. This would include, for example, a petrol filling station.

Explosive atmospheres are also found in places like distilleries, paint spraying plants, flourmills, woodworking machine plants, and the water industry.

Failure to ensure safe working practices in these circumstances could result in the ignition of explosive gases, or dust clouds, leading to injury or even fatalities. Clearly, sign company electricians working in such environments need to be appropriately qualified. It would be folly to employ electricians who were under qualified. The qualifications for work on petrol forecourts are Comp'Ex 07 and Comp'Ex 08.

Comp'Ex 07 relates to the preparation, installation, and decommissioning of electrical installations at petrol filling stations.

Comp'Ex 08 relates to the inspection, test, and maintenance of electrical installations at filling stations.

You can find out more about Comp'Ex 07 and 08 at www.compex.org.uk.

8. **Tell me about the CSCS (Construction Skills Certification Scheme) qualifications relating to your workers, supervisors, and your company?**

The most basic level of CSCS qualification – which every site-based employee must have, without fail – is the CSCS Construction Site Operative Green Card. There is, quite simply, no excuse for having on-site employees who do not possess this extremely basic qualification.

This is available via NVQ level one or by employer's recommendation, so it really is not very challenging to achieve, but at least (one hopes) it will help to do something to focus attention on health and safety at its most basic level.

In the sign industry, construction skills are important. Over the years, we've seen people with poor construction skills push health and safety to the limits. For example, heavy signs erected at motor dealerships and petrol forecourts, with dangerously shallow foundations, standing against the wind like sails.

A MID sign at a filling station (the upright sign containing the prices), weighs up to two tonnes, and may be in danger of toppling over into a public area if the foundations are not properly constructed. The same may apply to large signs at superstores, and the entrances to out-of-town shopping malls. You can find out more about CSCS at www.cscs.uk.com.

9. Does your Project Manager have the necessary training, and experience, to comply with The Construction (Design and Management) Regulations, 1994, known as the CDM Regulations, on your sites (these regulations are due to be updated in 2007)?

The CDM Regulations offer protection to your company until a building, or its sign, is demolished, and all the remaining materials are disposed of. You can find out more about the CDM regulations at www.hse.gov.uk.

10. Have any of your managers completed the training required of crews?

Best practice would have it that senior managers, as well as their crews, complete courses like the UK PIA (Petroleum Industry Association) Forecourt Contractor Safety Passport Scheme. In addition, there is no reason why all employees should not complete a behaviour-based safety course of some sort.

Doing the latter sends a loud signal that the company *genuinely* regards health and safety as a critical issue, and reduces any pressure by managers on crews to cut corners on commercial grounds.

11. Does your company have a good grasp of the Working at Height Regulations, 2005?

If not, and the project requires working at height, they are putting you – the customer – at a potential risk of criminal prosecution, or civil action. Test their understanding of the regulations; ask if their employees have been fully trained and, if so, how.

Ask them whether they insist on a Working at Height checklist. A checklist not only gives you an auditable trail demonstrating compliance with these regulations, it also shows that the company has a mechanism for constantly highlighting their importance with employees. Ask to see a copy of their checklist. You can find out more about the Working at Height Regulations at www.opsi.gov.uk (or just enter "working at height regulations" into Google).

12. Have your crew members attended a behaviour-based safety programme (such as the safety-in-the-mind Incident and Injury Free course)?

It is not a "must", but it certainly shows focus on health and safety issues, which is in your interests as much as the sign company's. A safety culture means fewer risks, fewer injuries, and less litigation; and good safety procedures reduce cost in the long term.

13. **If the company claims to do electrical work, is it enrolled as an approved contractor with the NICEIC (National Inspection Council for Electrical Installation Contracting)?**

If work is carried out on electrical circuits, and the company you employed is not enrolled with the NICEIC, you may be on dodgy ground. In the event of an accident, investigators may well ask you, personally, why you authorised a company to carry out electrical work for you, when there was no proof that they were qualified to do it. You have a responsibility to choose a competent contractor. You can find out more about NICEIC at www.niceic.org.uk.

14. **Are all your electricians accredited to City and Guild qualifications C&G2380/2381, and your test and inspection engineers accredited to C&G2391?**

If not, are they qualified to do the work? Don't let the contractor allow unsupervised 'improvers' (basically, trainees) to carry out electrical work on site. C&G2381 is the Certificate in the Requirements for Electrical Installation, which helps you to comply with BS7671: 2001, and covers topics such as "protection for safety, selection and erection of equipment, and inspection testing to meet the standards of IEE Wiring Regulations". This qualification succeeded C&G2380, but if your sign company's electricians have C&G2380, that is absolutely fine. These qualifications are sometimes referred to as "16th Edition" courses. C&G2391 is the Certificate in Inspection, Testing, Design and Certification of Electrical Installations, and "satisfies the NICEIC's requirement that prospective qualifying managers and operatives fully understand, and comply with, the requirements of BS7671 for inspection, testing and certification of electrical installations." It is entirely acceptable for you to ask to speak to the sign company's qualifying supervisor, personally (they will only have one if they are an NICEIC accredited company). You can find out more about these qualifications at www.city-and-guilds.co.uk.

15. **Is all electrical work done to meet the Institute of Engineering and Technology's 16th Edition Wiring Regulations (BS7671:2001, incorporating amendment 2:2004)?**

No ifs or buts on this one: either it is, or it isn't. If it isn't, then do not appoint them! Incorrect wiring is one of the most common causes of fire, but quite apart from this, often results in illuminated signs failing, or simply looking awful. If a sign 'trips out', it can disable an entire site – even to the point of preventing fuel being dispensed, if it happened in a forecourt.

Testing for faults, and issuing of certificates, should *only* be carried out by qualified electricians. You can find out more about these quality standards at www.iee.org

16. **Where does your sign engineer's remit stop, and the electrician's remit begin? Do you have clear procedures to cover this?**

Check that sign fitters are not doing electrical connections that they are not qualified to do – potentially, *you* are ultimately responsible in the event of a tragedy.

Electrical work should be done by appropriately qualified electricians. Check this is clearly understood, not just by the salesman sitting in front of you, but by the sign engineers and electricians that will carry out your work. You need to see evidence of systems to ensure that all employees adhere to the rules.

17. **Are all electrical components, and signs, tested to the standards laid down in EMC (Electro Magnetic Compatibility) Regulations, where applicable? Are they clearly identified with CE approval marks, as specified by the Department of Trade and Industry, and as required by law?**

If they are not, you could be served with a removal notice for non-compliant signware, for which you (not the sign company) will be deemed responsible. You may be able to put the onus of replacement on the sign company, but only if they are still in business, and almost certainly, only with a great deal of hassle. You can find out more about EMC at www.theit.org, and more about CE marks at www.bsi-global.com.

18. **Is your sign company accredited to the International Contractor's Safety Rating System, and if so, what was your score?**

Sign manufacture, installation, and maintenance involves many hazardous activities, and responsible companies will attempt to reduce the risks. This accreditation commits the sign company to an independent annual audit, to ensure compliance, and protect the customer against risk or loss. I would suggest that a minimum rating of 2,000 points is acceptable.

19. **Is your purchasing manager a Member of the Chartered Institute of Purchasing and Supply (CIPS), and will he/she work continuously with suppliers to deliver better value?**

A professionally qualified purchasing manager should be able to increase value, and deliver lower cost, on most projects, over time.

Most sign companies are able to run a project more cost-effectively 12 months after they have started work, than they can at the beginning. This is because they develop specific knowledge about that particular contract, and learn how to do things more effectively. Do they have a formal means of capturing that information, applying it, and sharing the benefits with you, or not? You can find out more about CIPS at www.cips.org.

20. **Can you prove that you have worked with large customers on cost transparency programmes?**

They should be able to show you that they have saved considerable sums - or delivered added value (perhaps as much as 20% of the tendered project cost) - by working to make operational cost savings. You can't deliver gainshare without transparency (see question 21).

21. Can you prove that you have worked with large customers to deliver gainshare?

Gainshare is one of the few methods of financially motivating a supplier to deliver reduced cost for a customer. It means, in essence, that supplier cost-savings will be shared with the customer. Whilst many manufacturers will understand this, and will have made it work for others, it is worrying that some senior managers in the sign industry *still* don't know what 'gainshare' actually is, and haven't bothered to find out. Don't let them squirm out of this – the sign industry absolutely lends itself to gainshare initiatives, and they can offer huge financial benefits, to both parties. You can't offer gainshare without cost transparency, by the way.

22. Have you established partnerships, with your customers, to share good practice; and has that been used to benefit all parties?

If their answer is yes, ask for proof, and documentary evidence of benefits. Ring customers and validate the claims.

To give one example from my own experience: on one project, we discovered that a sign used repeatedly, on all of a customer's premises, was being manufactured by nine different companies. Each had followed the technical specification, but they had used minor differences in fabrication and dimensions. Initially, this was fine – the signs all looked "right"; but when some of them were damaged, later in life, it was discovered that some of the original manufacturers were no longer in business (a typical situation in the sign industry), replacement panels did not always fit, and no companies had the tooling to make the correctly-fitting panels. We designed a survey document, which helped us to identify all the key differences; that gave us the information we needed to design a universal panel kit, which could be adapted to fit eight of the sign types. The information was shared with the customer, so that other maintenance companies (yes, even our competitors!) who also worked for the same customer, were more likely to get it 'right first time', saving them (and the customer) time and money. It wasn't the most aggressively competitive stance to take, but it was the most professional, and the customer appreciated that and saw the financial benefit of our expertise.

23. Can you give me the name of any blue chip referees, who can talk about increased value, better service, cost transparency, and gainshare?

There is no acceptable answer to this apart from 'yes, here are their names and telephone numbers'. Ask them about contract values and value creation - anything else is waffle.

24. Has the company received any awards?

Any kind of business award gives some element of credibility; if they are as good as they say they are, they will surely have won *something?* The best awards are those given by the customers themselves. Awards show external validation of the claims that the company is making about itself. They should have got *somewhere,* in *some* awards, surely?

Don't be *too* taken in by this, however, because the judging on some awards schemes is perfunctory to say the least (I know of some companies, winning awards for commercial activities, that had County Court Judgements against them, which rather nullifies the kudos endowed by the award, and which was, apparently, missed – or, worse, disregarded – by the judges).

25. Has the company received any compliments from its customers?

Don't believe them until you've seen evidence. Customer expressions of gratitude, thanks and appreciation, are not really part of the culture of the industry, but nevertheless, the best firms *will* have them, so ask to see examples.

26. Can I get access to job details on-line, during the life of the project?

There should be no reason why you shouldn't be able to get access to your estate list on a database, which should include work carried out, job status, and associated pictures, via a secure link on the supplier's website. This is a quick and convenient means of checking on job progress, viewing potential problem areas etc.

27. Does your company have a dedicated value engineering and design team?

The search for value should not just be something that is talked about, but should be built-in to the structure, and culture, of the company, so seek evidence that this is the case.

28. Has the company got a dedicated cost reduction team?

It should have one. If it does, there should be representation from Purchasing, Production, Installation and R & D (Research and Development), each with clearly defined targets.

Some sign firms are reluctant to think like this, and fear that if they reduce costs each year, customers may *believe* that they were, previously, being fleeced. This may be about management fear, and a lack of mutual understanding, as much as it is about reality; the company may very well *not* have been overcharging. Most customers, however, are far more sophisticated than this, and understand that cost-reduction, or value creation, is possible, over time, on a project, so the problem shouldn't arise. That doesn't stop sign company managers fearing that it will, and being resistant to gainshare, though.

Gainshare should be an integral part of any long-running project, and a formal process for securing cost reduction goals should usually be written into the contract.

SAFETY, QUALITY, TRICKS AND LIES

29. Has the company succeeded in reducing costs, for any other customers – and can they show you evidence of this?

Serial manufacturing of identical signs *always* offers up cost reduction opportunities. If they can't show you evidence of this, they have either never offered cost reduction, never worked on a project with sufficient critical mass to deliver cost reduction, or they are unsure of the process altogether. None of these scenarios are good. If, after a year or so, a serious company can't deliver cost savings, or added value, on repetitive work, they just aren't trying (or are keeping the cost savings to themselves).

30. Can you prove that you have handled large roll-outs?

If your project involves the manufacture, and installation, of signs across the UK, in a tight programme to complement, for example, television and print advertising, then you need assurances that the company you appoint has done it before. Don't let *your* project be their first.

31. Can you prove your ability to handle fast prototype projects?

Obviously, this is one of those questions which is not necessarily applicable to everyone. If it is, check whether or not they are able to give you 'intelligent reporting' on the installation and maintenance of a prototype, so that you know, objectively, how they will perform, before ordering a national roll-out.

CHAPTER ONE - CREDENTIALS

You need good calibre, accurate, technically competent reporting back to the manufacturing base. This will tell you if (for example) one component has a poor fit, before you, or the sign company, order serial manufacture of the component. Unsophisticated companies, unused to this scenario, may fear criticism if they report a component needs adjustment at the manufacturing stage – yet that is precisely what prototype projects are for!

32. Can you prove your ability to work with new materials?

There is a huge array of new materials coming on to the market; some are lightweight, short-term graphics, others are more durable, involving advanced technology. You need to be sure that the company has the expertise and skills to manage various materials effectively, and to take advantage of newly available materials to offer the best solution for your particular requirements.

Chapter two

Financials and Sub-Contracting

The UK sign industry is highly fragmented. Most of the work is done by relatively small companies, often in existence for less than a decade. Whilst there *are* many older sign companies, in general, there is a relatively high turnover of firms.

The standards of formal management, technical skills, quality management, and health and safety, are often below par when compared with most other sectors of industry and commerce. At a basic level, the entry threshold is low – you can set up and call yourself a "sign company" with only a small amount of capital.

This leaves the industry accessible to some individuals who wish to run their own companies, but who find other business sectors beyond their reach, capabilities, or skills base. Some sign companies' business models are crude, and short-term; tactical (rather than strategic) management is the norm.

To add to the difficulties, many of these companies lead a cyclical life where (rather like the military), there are long periods of relative inactivity, followed by short bursts of *intense* activity (whenever they win a large project).

Larger contracts will often require significant up-front capital, and yet, often, large customers take a long time to pay. Managing this calls for a certain level of financial competence, particularly in relation to wildly fluctuating cash-flow; conventional management disciplines would demand very high levels of forecasting accuracy, superb contract management, skilled cash management, and diligent invoice management, in the face of such a situation; regrettably, these skills are not often in evidence in the sign sector.

Large customers, who pay late, create a specific set of challenges for the management of any supplier company. Coping with these difficulties is quite beyond many sign companies, however, and company failure, due to cash-flow difficulties and overtrading, is common.

The ecstasy of a large order is frequently followed by the misery of company collapse: the sign firm has to pay up-front for materials, and ensure that it has the resources to complete the project, and it will usually do this with borrowed money (few sign firms are well capitalised); contract delays, and sluggish payment processes, will then force the collapse.

In the company's dying days, it may attempt to stave off the inevitable, by completing some sites hurriedly, and placing increasingly strident, panicky calls to the customer, demanding money. Hurried completion may also have involved the cutting of corners.

The brutal truth is that Sharon in Accounts at Big Company PLC leaves the office at 5pm on a Friday, regardless, and is blithely

CHAPTER TWO – FINANCIALS AND SUB-CONTRACTING

unaware of the fact that unless some poor sign company receives its cheque on Monday morning, the bank will move in, and 30 people will lose their jobs.

The bureaucracy, checks and balances, and financial controls, built-in to large company invoice payment are sometimes a complete anathema to sign company managers, who, for the most part, have never experienced corporate life at a senior level. But this is just life in big companies, and suppliers must be robust enough to weather it – if they are not, they are unlikely to survive.

For the customer, particularly customers with large national roll-outs to place, this poses some clear difficulties. The failure of a supplier company in the middle of a time-critical project is an obvious disaster, and the solution will usually lead to substantial, unbudgeted costs; but such situations happen every year.

Clearly, this is all generalisation, and like all generalisations, there are exceptions – many of them. A brief look at the company's financials before you begin will tell you a great deal.

33. What is the company's credit history like?

Never, ever appoint a sign company without doing a credit-check – you should ask the question, but *never* just take their word for it. The industry is rife with financially unstable companies, often sporting County Court Judgements (see question 37).

Many of the top sign companies 15 years ago are no longer in business – it is an *appalling* record. Looking at last year's profits won't tell you whether they are running out of cash, or at risk of overtrading, so do the credit checks – there are plenty of good, cheap systems to help you on the market.

For further information on how to do credit checks, go to www.creditsafeuk.com.

34. Is the company profitable?

Ask the company, but also, check yourself. You'd be surprised at how many companies in the sign industry aren't.

Every year, a lot of new sign companies start up, and many cease trading, for one reason or another – the rate of churning is high. This is a sure sign that many business owners simply 'have a go' at running a sign firm, because they think it will be easy, and then fail.

Profitability is very different to credit history, and you should check both. Ask yourself how disruptive, and costly, it would be, if your sign company ceased trading either immediately before, or at a critical point during, a new store launch or rebranding programme. Ask yourself what the overall cost would be to retender, appoint, and brief a new supplier – let alone the time this would take.

CHAPTER TWO – FINANCIALS AND SUB-CONTRACTING

35. Can you demonstrate what percentage of your turnover represents sub contracted services?

It is only acceptable if the sign company's sub-contracting is less than 15 per cent of its turnover, simply because, otherwise, you would be paying double mark-ups. Prices, and quality standards, should be in the hands of your supplier, not your supplier's supplier.

If the company has a large sub-contract percentage, because (for example) it sub-contracts *all* installation and maintenance work, it is wise to seek firm, credible and *provable* reassurances of the quality of those companies. Sub-contracting itself is fine in principle, so long as you are confident about the sub-contractor, and have been satisfied as to their qualifications and quality procedures (you should *not* be satisfied with a verbal confirmation from the main contractor, by the way).

36. How do you appoint, and measure, sub-contractors?

If sub-contractors *are* going to handle a significant element of the contract, it might also be prudent to do credit checks on *them*, which is why this section is in the chapter on 'financials'. You can't do checks on them, however, if you don't know who they are.

The sub-contractor's performance has a direct bearing on the quality of the work, and, potentially, on your health and safety record. Ask yourself what could happen if they were not insured? Bear in mind that a two-man business doesn't legally need a health and safety policy (many sub-contractors are two-man firms). Ask yourself what would happen to you, if one of the two was killed whilst working on your premises?

Some sign companies will be precious about giving you details of their sub-contractors; their supposed concern is that you might go direct to the sub-contractor in future, and cut them out. If they are so lacking in confidence that this fear is legitimate, then you might draw some negative conclusions from that. The real reason will probably be that they are making a margin.

This is fair enough – after all, you are using their contacts, and they have taken the lead role in the contract, and spent time negotiating with the sub-contractor. But you should have transparency about that margin: the usual reason for being coy about it, is that it is excessive. 10-15% is perfectly reasonable, and if the margins are at that level, you should pay them without quibble, all other things being equal. It would probably cost you more than that to handle multiple projects directly, and (so long as they are doing it competently), they will do a better job, because they are used to doing it. The proviso is, of course, *'so long as they are doing it competently.'*

37. Has your company got any County Court Judgements?

These are rife in the sign industry. Some sign company executives even believe that having them is the 'norm', and are, apparently, completely unaware that it is not necessarily the 'norm' in other industries! Don't take their word for it, check; but always ask, as well. I know of situations where sales people have lied about this; this is fairly surprising behaviour, because it implies that they do not have sufficient business 'nous' to realise that this information is in the public domain. Or, perhaps, they think you'll never check. But if they lie to you, then you know two things: firstly, they are dishonest, and secondly, their financial stability deserves further investigation.

This is not a black and white issue; even large, reputable companies may have one or more County Court Judgements, perhaps because they are slow payers. It isn't a cause for celebration, and they exist for a reason, but discovery of a CCJ calls for further investigation, and intelligent interpretation, not knee jerk reactions.

Chapter three

The Contract

When problems occur between sign companies and their customers, the contract is often at fault. Problems usually arise because of the difference between expectation and delivery.

Many sign companies are *dreadful* at contract negotiations, and frequently see them as an *obstacle* to sales. For them, it is a "difficult area", to skip past quickly, before anyone notices.

Often, this is because so many sign company managers, and salespeople, lack the basic business training that would enable them to have a mutually useful contractual discussion with a customer. However, if they can't do something as simple as that, can they really manage a sign roll-out involving complex project management, and negotiations with sub-contractors and their own suppliers?

It has to be said that a properly drawn-up contract, with clear Service Level Agreements, will remove 'wriggle room' on *both* sides of the agreement. If you are entering into the agreement in good faith, you should insist on a proper contract, because it will help to protect you against any shortcomings by the sign company. By the same token, it will also provide them with a level of protection.

Make sure the contract covers all the details. For example, prototyping – testing new designs and material to ensure that they perform as expected (this might, typically, involve testing colour fastness, or ensuring that performance is good during both daylight and night time conditions).

It should also, nowadays, cover lifecycle costs. Disposal costs are significant – you can't just hurl old signs in the rubbish and expect the council to take them away – but they are sometimes overlooked. For your own legal protection, make sure that old materials are disposed of properly.

Finally, what happens when things go wrong? The contract should specify some sort of dispute procedure. Guarantees, and warranties, will be separate to the contract, of course, but make sure you have them. If you are using several contractors, make sure that you are clear about the areas of responsibility between the manufacturer of the materials, the manufacturer of the signs, the installer and maintainer of the signs, and your own company as the user. All too often, complaints get bogged down in dispute and counter dispute between contractors and sub contractors, and when that happens, the customer is likely to be the loser.

38. Will you agree to a confidentiality clause?

With such a clause, you can have some confidence about sharing sensitive information with them, which may help them do a better job for you. Make sure, however, that the agreement is binding on all employees, contractors *and* sub-contractors hired by the company, and that the onus for enforcing confidentiality is on the sign company.

Having a confidentiality agreement protects information which you would regard as commercially sensitive, and this is the main reason for having one. It also covers you in a multitude of other situations; to take one example, if their crews accidentally discover something you'd rather they didn't know, it may dissuade them from gossiping about it, or running off to the media with a story. There is no reason *not* to insist on a confidentiality clause.

39. How will you ensure that your *sub-contractors* don't gossip about my project, to my competitors, to the press, or others?

The confidentiality clause should also apply to sub-contractors. Ask for written proof that *all* sub-contractors have signed a confidentiality clause, in their standard contract with the sign company, and that they understand it. The clause should specifically forbid the sub-contractors talking to the press, in the event of an accident; for example, not naming and blaming an oil company for a fire which is covered on national TV.

Another common reason for confidentiality is that signs relate to a campaign, or a new brand, which is being unveiled at a specific time, as part of a sophisticated nation-wide (or global) marketing strategy, linked to TV and print advertising. In this event, it would be unhelpful to have sign company employees blabbing about it beforehand.

40. Is the contract "off the shelf", or specific to this project?

Frequently, if they use one at all, sign companies use "off the shelf" contracts, which bear little relationship to the work they are being asked to undertake. This may be because they are hoping to drive a coach and horses through the project, and apply hidden charges; because they won't invest in having a professional contract for each project; or because preparing a specific contract for each new project is, frankly, outside their skills envelope.

A professional, project-specific contract will enshrine both parties' expectations, and service deliverables, so it is in your interest to insist on one.

If producing a proper contract is beyond them (which it quite often will be), it is worth investing in having one done for them. This might be unusual, in that you are the customer, but that's the nature of the sign industry, and having one done professionally will almost certainly save you money and hassle in the long run.

41. Will the contract use Key Performance Indicators (KPIs) and Service Level Agreements (SLAs)?

If not, what are you expecting of them, and what if they fall short of this? You must build-in to the contract a means of assessing how well the project is going. If you are dealing with a quality-led company, they will be very receptive to this. They may even suggest it themselves, because they know that their high performance level will be objectively measured by this system, and if they are offering a high level of service, they will want to be able to prove it to you!

Service Level Agreements are a routine part of contract negotiations in most other industries, and there is no reason at all why this should not be true in the sign industry.

Contractual Service Level Agreements and Key Performance Indicators remove many of the things which may, later, become niggles and causes of discord.

Be realistic about this, however. Let's say that you employ a sign company to cover signs across the entire UK, and the contract includes a four-hour call-out option for emergencies – in other words, the sign company agrees that it will get crews to any location in the UK, within four hours of a call-out. If, at year's end, they have scored 98%, or above, in reaching this target, that's fair enough; if they have given good service, generally, across the scope of the contract, and yet you penalise them for the remaining 2%, you are being unreasonable.

Regardless of what it says in a contract, no employer can absolutely guarantee that, on any given day, it will not have employees who are sick, stranded by traffic congestion or accidents, overwhelmed by the consequences of natural disasters etc. Every company in the world employs people, and ultimately, people are fallible – however good their intentions. Being petty about this simply encourages them to attempt to deceive you.

On which point: if they claim 100%, on a large contract, you should be pretty suspicious. What is the reporting procedure, and is it open to abuse?

Chapter four

Project Management and Contract Implementation

The company's project management capability will determine whether or not the job goes well. You need to ascertain whether management is system-based, and rooted in good training, or whether it is of the make-it-up-as-you-go-along variety.

Project managers should have the proper software and hardware, as well as an infrastructure, which supports a good command-and-control system, with accurate feedback. Here are some questions to help you find this out.

42. Do *all* installation teams carry a digital camera?

There is no reason why they shouldn't. Cameras are cheap, easy to use, and make reporting back on problems easier, and more accurate. Let's face it, not all sign installation and maintenance crews are hired for their command of the English language, even in the most professional companies. Narrative descriptions often suffer from excess brevity, lack of clarity, and poor English. In these instances, it is always helpful to have a supporting picture.

43. Have you got good project management disciplines and procedures?

This will help you avoid all sorts of petty contractual disputes. Many of the problems which occur between the customer, and the sign company, centre on service delivery. Good project management software will tell you exactly what has happened, when, and where.

It will also, incidentally, tell you how long *your* management people took to respond to requests for approvals or authorisation, so that you can see how your part of the contract is going.

Have a relationship of trust, with a professional contractor, and encourage good, accurate information. It will help both you and them. There is no excuse, in this day and age, for not providing a comprehensive, transparent project management system, available whenever you need it.

44. Have you got the capability to provide a comprehensive software-based asset register of signware?

A sign company should be able to provide you with the usual updates, via *your* chosen method of communication. You should receive supporting photographs and a running update of ongoing work. You should have secure access to this via the internet.

CHAPTER FOUR – PROJECT MANAGEMENT AND CONTRACT IMPLEMENTATION

45. Do you provide "intelligent reporting" of on-site problems?

This is one of the most common issues in sign installation. Typically, an installation company will not be paid until the job is signed-off, or completed. So, when they come across a component that doesn't fit, the last thing they *want* to do is report it. If they do, it will go back for re-manufacture, the whole project will be held up, their work schedule will be disrupted, and their payment will be delayed. Ethical companies will report it, regardless; others won't !

Their solution? If the component is "out of sight" (perhaps high off the ground, or an internal part of a sign not visible from the outside), some will apply hammer force until the component *does* fit. The job will be signed-off, the company will be paid, and by the time the component fails (which will be well before the end of its design life) any action from the customer will be too late. The same failing may occur at your sites all over the country, and you will have to pay for multiple site visits to put it right again. Warranty claims will be bogged down as contractors, manufacturers, and maintenance companies all blame each other.

This happens a lot, because sign components are often originals, made for the first time by small companies, with limited resources. "Intelligent reporting" allows for comprehensible technical reports, which can be delivered as a matter of urgency, with supporting digital photography. The reports highlight problems in their early stages; the faulty component can then be remanufactured, and the problem fixed during the prototype project (or development phase), at much less expense. You are more likely to get "intelligent reporting" from a company with a lot of installation engineers, because their critical mass and advanced project management capability means they can cope more easily with the disruption, and are more likely to place a high priority on good, long term relations with the customer.

46. How will the success of the project be measured?

In addition to sound project management, and a good contract, you need a mechanism for measuring Key Performance Data as you go along. This will enable you to spot problems early, and negotiate penalty clauses from a strong position. Insist on KPIs (Key Performance Indicators) and SLAs (Service Level Agreements).

47. Does your company actively operate a continuous improvement programme throughout its business?

You should be shown evidence of a formal system, throughout the company, for capturing information about lessons learned. You should be looking at documentation which includes graphs and charts, not just listening to a meaningless piece of Ricky Gervais-style management jargon.

48. Does your company carry out regular supplier audits of its sign manufacturers, component manufacturers, or suppliers, (as appropriate) to protect the quality of its installations, and to monitor the quality of components used?

CHAPTER FOUR – PROJECT MANAGEMENT AND CONTRACT IMPLEMENTATION

Your brand's new signs won't look new for long if the installer uses cheap, de-specced components to get the costs down, and win a tight tender, or improve the margins. Companies claiming to offer a premium product should have *total* control over their supply chain – otherwise, you should take control yourself, and hire other companies directly.

49. What expectation should I have of your "making good"? I would like a *detailed* answer, please.

Most sign companies will paint out all screw heads, and sweep-up after themselves, but check that they will leave the work area to the standard it was in before they started.

Make sure they have allowed, in their budget, for any re-turfing, re-painting, re-concreting and cleaning costs, which might be needed after the job is completed. It is much more difficult, and expensive, to get them to re-visit. Excluding these costs may be a way to further undercut competitors; if you don't *want* any making good (perhaps because you can do it internally, more cost effectively), then say so as part of the tender, so that all the companies pitching for your business do so on a level playing field.

50. Can you demonstrate improved service delivery by showing me your Key Performance Data?

Again, if they haven't done this for previous contracts, they are not going to do it for you. Managers need to be trained to do this professionally, although any company performing the same service, for any length of time, will surely be able to do a better job, more efficiently each time? If they can't show you their performance data, they just aren't trying. If they don't understand the question, they probably have *no* Key Performance Data.

51. Are your project managers always available by mobile phone and/or email?

On a complex roll-out, you have every right to be able to contact project managers, at any time convenient to you, with the exception of very unusual circumstances.

52. Is there more than one dedicated out-of-hours contact number given in case of emergency?

If you only have one emergency number, what is going to happen if there is poor reception, if they're engaged, or if they sleep through your call? 24/7 should be just that!

53. Have your crew vehicles got tracker devices fitted, so that your call centre can swiftly locate the nearest vehicle to an emergency call-out? And have they got satellite navigation?

Again, across a fleet, tracker devices are relatively cheap, and easy to fit. Any credible organisation will have them. It is depressing that even after many years of becoming commonplace in professional fleet

CHAPTER FOUR – PROJECT MANAGEMENT AND CONTRACT IMPLEMENTATION

management, many sign company managers regard tracker devices as exotic "new technology".

Time wasted by the crew wandering around looking for a site in the countryside could well cost *you* money. Satellite navigation is a cheap and easily available solution, particularly when purchased in bulk for a fleet, so there is no reason for them not to have it.

54. How can I check on the progress of a job, at a particular site, at any one time?

The best companies will instruct their teams to take digital photographs of the site as they leave it; the photograph will be emailed to the sign company, and then on to the customer; or directly to the customer; or posted on a special secure website, which is constantly updated, and to which the customer has security-controlled access. In addition to the picture, the website will contain detailed information about progress at sites across the breadth of the contract.

This is by far the most cost-effective solution, for both the customer and the sign company, but the major benefit to the customer is that it gives real, up-to-the-minute information, about the state of the contract, at any stage.

Chapter five

Health and Safety

In today's business culture of corporate manslaughter, no-win-no-fee lawyers, compensation, and increasingly stringent health and safety legislation, the onus is (unfortunately) on *you*, the customer, to ensure that any work carried out on your sites meets, at the very least, the minimum standards laid down in the law.

If it doesn't, you may find your insurance invalidated, and your ability to defend any subsequent court case impaired. There are plenty of company directors, and senior managers, who are bankrupt, or serving prison sentences, because they took this less seriously than they should have done. The legal consequences may, in certain circumstances, fall on you, as an *individual*, not just your company.

Having said all that, the law is not out to get people who make a simple mistake, or an oversight. We all do that, all the time, because we're all human. It is designed to deal with those who regard health and safety as an obstacle to business, and something which can be circumvented with a little bit of deviousness.

Unfortunately, you need to weed out the cowboys before they start putting you, and your business, at risk of criminal or legal proceedings. In most situations, even if there *is* an incident of some sort, strong evidence of diligent investigation of a supplier's approach to health and safety will weigh heavily in your favour with a court.

One health and safety expert states that 80% of people taking part in his training courses do not understand the nature of the risk when it comes to asbestos. In fact, asbestos-related disease kills 4,000 people a year, and that figure is predicted to rise, by some, to 10,000 deaths per year within the next decade.

I make no apology that this is the biggest chapter in the book, because the sign industry is a high risk sector. The chapter is about saving lives, preventing injury, and reducing the chances that you – and your company (not just the sign company) – will end up in court. Health and safety is of increasing importance, and is constantly subject to new legislation, civil law, case law, rules, and legal scrutiny and challenge (often through no-win, no-fee lawyers). You cannot leave it to the contractor to get it right – and you must make it clear to the contractor that flouting good practice will result in instant termination of the contract.

According to statistics from the Health and Safety Commission, 220 workers were killed in the year 2004/5 in work-related health and safety incidents; but we managed to kill a further 361 members of the public. There were around 363,000 'reportable' injuries, and more than 150,000 "other" injuries. Two million people believe they are suffering from an illness caused, or made worse by, their work.

Of the 35 million days lost to illness, 28 million were due to work-related ill-health, and seven million to workplace injury.

Of course, some industries will be worse than others. There are league tables of industrial sectors showing how they all fare, but the sign industry is not shown separately (it is usually filed under 'construction'). If it was shown separately, it would be pretty high up the table, in my estimation.

55. Is your company willing to tell me about its RIDDOR (Reporting of Injuries Diseases and Dangerous Occurrences Regulations) reportable accident reports?

It may have none, which is fine; it may have a small number, which, strangely, is *also* fine (unless the numbers are excessive). Accidents *will* happen; what is important, however, is that each accident has been professionally dealt with, and steps have been taken to ensure that all employees have learnt from previous mistakes. It doesn't have to be something related to the core business – remember, health and safety is holistic, and your approach to it should be comprehensive. In the case of my own company, for example, we had a RIDDOR reportable accident relating to an employee with an existing bad back, who aggravated it whilst changing a van wheel.

You must convince yourself that the reporting and investigations are being professionally carried out (and that information about accidents is not simply being suppressed, using an internal bullying culture). This is particularly important in companies which claim to have had no RIDDOR-reportable accidents. That is possible, particularly for relatively new, small companies, but in the real world, is increasingly unlikely.

56. Will your company pro-actively provide me with a monthly accident report, and record all incidents which occur during the project?

A good company will be transparent about its entire accident record. Sign installation often involves hazardous work, and on some sign projects, there *will* be incidents. When incidents occur, you need the facts, quickly.

57. Does your company routinely prepare a job-specific Risk Assessment, *and* Safety Method Statement, at *every* site, and for *every* job?

This is not just bureaucracy; it could, in extremis, help to keep you out of jail, given the severity of corporate manslaughter and other legislation today.

58. Does the company keep its training up to date?

Yes, "on the job training" is sometimes acceptable, and all companies do do it, but it can also be code for "we don't bother". Ask what their training commitment is, and whether, for example, a training needs analysis takes place for each employee.

CHAPTER FIVE – HEALTH AND SAFETY

59. On average, last year, how much paid time did each employee spend on training?

Some of the 'bottom feeder' companies in the industry may (privately) be thinking 'none of your business' if you ask this question, so watch the body language and try to detect that. It is most *definitely* your business, if there is any possibility of you hiring them; and if they are coy about answering, you are perfectly entitled to draw negative conclusions. In this context, "training" means attending courses run by credible external training organisations which charge a fee. *All* companies do on-the-job training (or at least, it is to be *hoped* that they do), but since you can't really measure that, the only meaningful measure is the amount of time and resource spent on approved external training courses. Can they demonstrate this to your satisfaction?

60. Are your company's employees rewarded for good health and safety practice? If so, how? Details, please.

The best firms give recognition, bonuses, or rewards, to their employees, to encourage good health and safety practice.

61. What are your specific health and safety targets?

Engage in the detail here, to see how committed your supplier really is, and to check that they aren't just paying lip service to fake targets.

62. Is a Working at Height checklist always used to record compliance with Working at Height Regulations, 2005?

In 2004/5, HSE statistics show that in the UK alone, 53 people died, and nearly 3,800 suffered serious injuries, as a result of a fall from height in the workplace. Further information on the Working at Height Regulations can be viewed at www.opsi.gov.uk.

63. If a worker on your site falls from height, and loses his life, or is seriously injured, can you prove that you provided a safe working environment for your employees, and appointed a competent contractor?

You can check whether they are up to the job by asking, for example, whether they use an anemometer to record wind speeds before ascending a scaffold tower. The Working at Height Regulations – which were amended in 2005 – need to be observed religiously, and the people working at height on your sites *must* understand them.

64. Who has audited your health and safety policy and can you prove this?

Take a good look at your supplier's health and safety policy. Does it look like an off-the-shelf package, or do they really understand it when they are quizzed in detail?

CHAPTER FIVE – HEALTH AND SAFETY

65. **Does your organisation's 'Health and Safety Policy and Procedures' document clearly define *all* responsibilities, and organisational rules, relating to potential risks arising from work activities?**

It's too late to start wondering who's at fault after an incident has happened. There should be a clear record of who is responsible for what. By being prepared, and knowing who is responsible, there is a reduced risk of incidents, as each individual is aware of the possible side-effects of their actions.

66. **Does your 'Health and Safety Policy and Procedures' document detail any alterations that may be needed to leadership and administration roles and employee communication procedures, following promotions?**

All companies should have a clear policy on health and safety procedures, including what to do following promotions and job changes. Clear communication is essential at all times, and there should be no lapse following staff reshuffles, or new appointments at a company. Everything should be clearly defined, and all employees should be aware of who to contact for further information.

67. **Does your 'Health and Safety Policy and Procedures' document detail job planning and hiring?**

It should, but don't just look at the procedures in the document, ask questions to see if they are actually used. Make sure that health and safety is right at the forefront of hiring and job planning, and ask to see examples where this has been carried through.

68. Does your 'Health and Safety Policy and Procedures' document detail employee management and training?

How often is the document updated? Is it specific to each role in the company, or very general? A classic mistake is to take-on new employees, perhaps for slightly different roles, and not cover this in the document or internal training. Make sure it's not an off-the-shelf 'quick fix' document that is only made available for picky customers(!) *All* customers should be picky, to make sure the workforce on their job follows a planned, detailed, procedure.

69. Does your 'Health and Safety Policy and Procedures' document detail emergency preparedness, health control, and Personal Protective Equipment (PPE)?

Each person on site should be able to assess any 'emergency' situation, and deal with it. They may not be able to solve the problem immediately, but they should know how to approach it, and be able to find a route to a solution.

The procedures should detail Personal Protective Equipment, so that there is no room for error. PPE is clothing worn by a person at work, protecting him or her against one or more risks to his health or safety; for example, safety helmets, gloves, eye protection, high-visibility clothing, safety footwear, and safety harnesses. In certain situations,

CHAPTER FIVE – HEALTH AND SAFETY

the PPE required is highly specified, and your sign company should know, and understand, the rules and best practice.

The truth is that these rules are broken routinely in the industry. I accept that it is very difficult to police an entire workforce, particularly when many of them are working remotely on sites around the country, but it *is* reasonable to ask for evidence that the company does everything in its power to encourage good practice.

70. Does your 'Health and Safety Policy and Procedures' document detail planned inspections, analysis, and updates?

Ask what specific elements were improved, and how the improvement process works, and is monitored. The company should have a record of past inspections, which you can view, in addition to any which are planned. An established record of past inspections is an indicator of how seriously the company takes its health and safety, and gives a clue as to how long this has been a priority.

71. Are Risk Assessments, and Safety Method Statements (or Job Safety Assessments), reviewed regularly, and if so how often?

Check that there is proof of the changes, and updates. The information should be readily available, to all employees, even those working remotely.

72. Are your operatives trained to recognise, and deal with, asbestos, when they come across it?

Asbestos is the biggest killer in the modern workplace. If operatives are not trained properly, there could be a huge lawsuit hitting your desk in a few years time, especially if it wasn't identified in your Asbestos Register, and communicated in writing, in advance of the company undertaking the work. One single fibre of asbestos is enough to cause death through mesathelioma, up to 60 or more years after initial contact.

The number of deaths from mesathelioma was 153 in 1968, and is forecast to reach up to 2,450 per year between 2011 and 2015 (source: HSE website). Some estimates say that the numbers will reach 10,000 per year within the next decade. Litigation cases are expected to increase pro-rata.

Even drilling into a wall, particularly in a building erected in the middle to late 20th Century, is a risky thing to do, unless the crew has checked the Asbestos Register, and again, it is potentially *you*, and *your* company which is in the legal firing line if something goes wrong.

73. Are all of your operators 'appointed persons' for first aid?

This training could save the life of one of your colleagues, contractors or customers – what price does the sign company put on saving a life? Arguably, it is not necessary for every employee to have undergone first aid training, which may take a week or more, but in the best companies, all those working on site will have attended an 'appointed persons' course, which is much shorter, but gives the basics. If they have gone to the trouble of doing that, it may be a strong indicator of widespread good practice in other areas.

74. Does your company insist on doing medical checks on all its site employees before they commence work?

Checks should include pulmonary, musculo-skeletal, cardiovascular, hearing, vision and controlled substance use checks.

If this is not done, how do you know that the electrician they have just employed, for example, isn't colour blind, or that the person working on the scaffold tower doesn't suffer from vertigo? Check that they have complied with the Health and Safety at Work (HSAW) regulations.

75. Can I read the minutes from the last six months' health and safety committee meetings?

This is a good use of your time. Of course, you don't have to read them all in detail, but simply check that they do exist, and sample them. The company's willingness, and ability, to deliver them, is an indicator of probable widespread good practice. The contents of the minutes will tell you very quickly how seriously the company takes health and safety, and what initiatives are being undertaken which will, ultimately, improve safety on *your* sites.

76. Can you prove that your health and safety committee minutes are distributed, or communicated to *all* employees?

If they can't, it is likely that they are just paying lip service to the issue, and that health and safety stops in their boardroom. Every employee in a company has *some* responsibility for health and safety (at the very least, taking responsibility for their own wellbeing), and there is no reason at all why the minutes of the health and safety committee should not be made available to the entire company. This should be a transparent process.

77. Can you give examples of health and safety bulletins that have been given to your field operatives?

These are an important step in building employee awareness. Bulletins will draw attention to specific health and safety issues which have arisen, perhaps in other industries, other companies, or even other countries, and which may contain useful lessons. They may highlight new ideas, new thinking, new laws, or the results of relevant civil court proceedings.

Anything which the company becomes aware of, and which can assist with the health and safety of its employees (or, for that matter, anyone else) should be placed in the public domain, and should be easily accessible by employees. This is also a means of sending constant signals to employees that the senior managers take health and safety very seriously indeed.

CHAPTER FIVE – HEALTH AND SAFETY

78. Have you adopted a 'near miss' culture to identify problem areas before accidents occur?

Ask to see proof of near miss reports. Near misses can be as useful, in terms of lessons learned, as accidents, but sometimes, companies take the view that "nothing happened, and no one was hurt" so no action is required. This is a slipshod approach, and is very revealing about the firm's general attitude towards health and safety.

79. Have you properly investigated any accidents, incidents, or near misses?

Ask to see the detailed investigation reports, root cause analyses, and resultant recommendations of an investigation. Remember, their repeated mistakes could lead to injury, or loss of life, on your site.

If you are talking to a company of any size, and they claim not to have any investigation reports because they haven't had any accidents or near misses, you may be justified in expressing some scepticism. It is possible, but pretty unlikely, and it is *more* likely that the reporting is inadequate.

Some firms that *do* have such reports want to keep them secret, because they see them as an admission of failure, which implies an unsophisticated understanding of health and safety culture at a senior level.

Health and safety courses for senior managers usually stress the importance and significance, of openness and transparency. Health and safety is *not* a competitive issue; if a firm has learned a hard lesson, and is able to make its operations safer as a result, it should be willing to share that lesson with anyone – including competitors. To refuse to do so is a sign of some fairly cynical thinking.

80. Are your teams regularly subjected to on-site health & safety audits?

If not, how would you demonstrate to a court, if it was necessary, that you had made every effort to select a safe contractor? An audit trail is good evidence of a strong health and safety culture, if you end up in deep water! Expect the sign company to do this as a matter of routine.

81. Is there a 'refusal to work' procedure for your company's installers?

If the crew spotted something unsafe on your site, would they know how to report it, and would the company know what to do about it? You may not want to experience delays, but it is more important that you *don't* experience an accident investigation.

82. Have you audited the power tools on site? Do they all get Portable Appliance Tests (PAT)?

Insist on an audit trail, and Portable Appliance Tests, to protect you. You can find out more about Portable Appliance Tests at www.pat-testing.info.

CHAPTER FIVE – HEALTH AND SAFETY

83. Are all components of the sign up to British (or European) Standards?

The component parts - for example, the transformers - should be checked, to make sure they meet British and European Standards requirements, especially if the sign contains a cold cathode.

To give an example, my company once received a sign that contained a 15kV transformer, sent in good faith, by a foreign supplier. The maximum voltage allowed by European Standard (EN61050) is 10kV, and if we had breached this, we, and our customer, could have had a legal battle on our hands. We had a system for checking these things, and we prevented it slipping through – but a less diligent firm would simply have accepted it. More information on British Standards is available at www.bsi-global.com; for information on European Standards, go to www.cenelec.org.

84. Do you have proof that you conduct regular Toolbox Talks?

If you don't, how do you know that safety is being discussed at site level?

A Toolbox Talk programme greatly reduces the risk of accidents, occurrences of ill health, and environmental damage, in the construction industry (in this context, that includes sign firms).

The legal consequences of failure on your site, could come back to you personally, or your company. Continued specific training is one of the most effective means of defence against this. Ask for a list of their

Toolbox Talk topics, and proof that employees have taken part in the programmes.

The best firms will ask employees to sign every time they attend a Toolbox Talk, or receive information after such a talk - it gives them a record, showing a consistently dedicated attitude towards health and safety.

85. How are emergency procedures communicated?

Do the sign company's employees know where the fire extinguishers, and muster points are, on all the sites on which they are working? Is there a means of communicating this, or ensuring that they check this out soon after arrival on site? To find out, make sure you check their Safety Method Statements or Clearance Certificates.

86. Can I see your Control of Substances Hazardous to Health (COSHH) assessments?

Installation teams may well carry them, or they might be displayed on a factory wall, but ask the company directly whether they have made efforts to ensure that people genuinely *understand* them. COSHH statements are fundamental if workers are dealing with any chemicals, or substances, especially fuel or solvents of any kind. To find out more about COSHH, go to www.coshh-essentials.org.uk.

CHAPTER FIVE – HEALTH AND SAFETY

87. Are your teams trained to spot hazards?

Hazard awareness is not difficult to teach, or to understand, and there is no excuse for not having made some effort to *raise* awareness of hazards, and to refresh that process regularly.

88. Can I see the induction training records of all the people that will be working on the sites involved?

If not, how do you know whether or not they have been trained to the standards that you are paying for? Some companies regard induction as an opportunity loss, and something that is just an obstacle to making new people productive. But the supplier cannot possibly be in a position to assure you (credibly) that your work will be carried out to specified standards, unless a professional induction process has been conducted – particularly so, if they are hiring new employees to fulfil the contract.

89. What procedures are in place for the disposal of any hazardous waste?

Hot cathode (fluorescent) and cold cathode (neon) tubes are just a couple of the items now deemed to be hazardous waste. Ask for proof of disposal procedures, including paperwork, to check waste hasn't just been put in a landfill skip, or dumped. If it has, it could be traced back to your site, and in certain circumstances, legal action could be taken against *you*, not just the contractor. You can find out more about hazardous waste disposal at www.environment-agency.gov.uk.

90. Can you show me your job assessment for the provision of Personal Protective Equipment (PPE)?

Personal Protective Equipment is sometimes (wrongly) seen as an 'easy' way to reduce potential injury; whilst it is often essential, the first considerations should be risk elimination, reduction, isolation and control. Ask to see specific job assessments for the provision of PPE.

PPE requirements for sign installers are very specific, particularly in areas like fuel forecourts. It is not something that can just be left to 'common sense'; crews have to know the rules, and stick to them.

The same applies for factory environments, though it is more likely that installers will be the ones to put *you*, or your company, in legal peril, because they will be working on your sites.

91. If we have an urgent health and safety related issue on site, can you offer a call-out facility, within a specified time limit, 24 hours per day, 365 days per year, to anywhere on the UK mainland?

If they say they can, ask to see how many call-outs they have attended to, and what their average response time was. A four-hour call-out across the UK is feasible for a company which has true nationwide coverage (although you will need to be sensible in the case of extremely remote locations such as the Western Isles, the Scilly Isles, or the Outer Hebrides, where ferries, tides, and weather may cause delays which are beyond reasonable control. It also fair to make allowances for the sort of traffic congestion that makes it on to radio news, and strands thousands on motorways!)

92. Are your vehicles fitted with automatic van fire trace systems?

If not, you should ask them what they think the implications are of a burning van on your premises. Sign company crew vehicles, by their very nature, may carry flammable materials, and you should seek evidence that every precaution has been taken, before you consent to allow them on to your sites.

93. Have you ever made efforts to find out how your employees, and sub-contractors, feel about health and safety?

The very best firms will do professional 'climate studies' amongst their employees. After all, if you are trying to improve the attitude of the workforce on health and safety-related issues, you need to understand their *current* attitude, because that is your starting point.

There is a 70-question climate survey, available for very little money, from the Health and Safety Executive (HSE). Using this, you can identify trends within the company of employee's perceptions towards health and safety, and gather recommendations from the coalface by those doing the job. Further information can be found on the HSE website: www.hse.gov.uk.

94. How up-to-date is your PPE Issue Register?

Check the issue dates to make sure that PPE does its job, (in other words, that it protects the operator against harm). PPE needs to be 'managed' – you can't just issue some kit once, and then forget it. Some of it has a specified life expectancy, some of it needs to be serviced, and you need to constantly ensure that the PPE being used is relevant in the light of current legislation and best practice.

95. Do you insist on the use of harnesses in ALL mobile elevated work platforms?

Whilst this is, undoubtedly, not popular with the crews on-site, it is essential in reducing risk. Crews are human beings, and will sometimes resist things that they see as inconvenient, uncomfortable or a nuisance, and the only way round this is to make extensive, continuous efforts to point out the importance of it, and the fact that company health and safety policy is for their benefit.

Alas, for some sign companies, this seems to be too much trouble, and they cease to bother. If that happens, they are, potentially, putting you, the customer, in legal peril, if there is an accident. For your own peace of mind, you should reassure yourself that harnesses are being used on all elevated work platforms.

CHAPTER FIVE – HEALTH AND SAFETY

96. Would you show me the audits of all safety harnesses that may be used?

Are harnesses crumpled up in the corner of vans, unchecked until it's too late? Why have them if they might not work? They should be inspected on a regular basis, by the sign company. A firm which allows itself to be vulnerable to the accusation (from an injured installation engineer) that 'they didn't give me the right kit' or 'the harness wasn't in good condition' is also putting *you,* and your company, at potential exposure to legal peril or bad publicity.

97. What lengths are the lanyards?

You'd be surprised by how many are either too short, or so long that the person using them could hit the ground before their harness started working. Just dishing out a rope, any old rope, won't do. There has to have been some thought and planning put in to this, and the company ought to be able to give you some reassurance that it has done this.

98. Can you show me your harness register?

Harnesses must be registered and controlled – the sign company managers can't just hand them out, randomly, and keep their fingers crossed!

99. Are logical risk control procedures followed?

The risk control procedure hierarchy is as follows: eliminate, reduce, isolate, control, (ERIC), PPE, and then discipline. Check that they follow, and understand, this procedure.

100. Can you show me your Mobile Scaffold Tower Register?

Some mobile scaffold towers do not have parts that are safely interchangeable, so if they are, it could result in mechanical failure. Check that all parts used are registered to one tower only.

Chapter six

Dirty Tricks

Unfortunately, *some* (thankfully, not all) sign companies will always try to engage in dirty tricks. Asking a potential supplier the questions in this book will give you a good idea of their general credibility, making it easier to spot any deliberate attempts to deceive you. In this section, I've outlined some of the most common dirty tricks that are played.

I should repeat here, as I have said elsewhere, that there are good sign companies out there, and that even the most rogue-ish are unlikely to have *all* these tricks in their armoury.

Detailed, persistent questioning, and sound contracts, are the only real defence for the novice; experience will come in time, and in this industry, usually, at a cost.

Some of the dirty tricks outlined here amount to criminal fraud, others are less serious, but as a customer, you shouldn't be on the receiving end of *any* of them. They have *all* been used in the industry.

Leading fraud investigator Brian Healy says that most frauds are simple. Brian is the Operations Director of Haymarket Management Services, which provides specialist advice on fraud prevention, and can also help investigate fraud, recover assets, and bring perpetrators to justice. Although the UK sign industry is a tiddler, in comparison to manufacturing, construction and many other sectors, his firm is *very* familiar with it.

"For the most part, fraudsters do not use a lot of cunning, and often, it seems, they simply don't expect to be caught," he said. "Some of the frauds committed in the sign industry are at the very simple end of the scale – though the potential rewards, had the fraudster got away with it, can be significant."

Fixings

All sign companies buy fixings (screws, for example) in bulk; the disreputable ones then make big savings, by using fewer of them than they should, or by using cheap ferrous fixings, which can seize or cause rust streaks. Reputable companies will use plated fixings, which do not react with other metals.

Another example of cheating with fixings is using silicone, to fix panel gaps. It is the cheap, quick way of making a sign cosmetically acceptable, getting off the site, and getting paid, but it is not the long term solution. It is not durable, and there is a professional method of panel gap fixing which sign engineers from reputable companies will employ.

CHAPTER SIX – DIRTY TRICKS

Fluorescent tube illumination

Every professional sign company will know precisely how many lighting tubes to put in a sign of a given size; *un*professional sign companies won't have any idea! If there are too many the sign may get hot, and components will fail; too few, and you will get shading, or bright spots, which will make the sign look cheap.

It's not rocket science, but reducing the number of tubes can be a route to de-speccing the job, without the customer realising. Check the quotes carefully to ensure that they specify the same number of tubes for each sign.

Also, check that the tube *type* is carefully specified. Using tubes made by second-rate manufacturers can save the sign company money, although it is more than likely that the savings will not be passed on to you, and the sign will look tired or faded.

Note that the difference between 'cool white' and 'warm white' tubes is massive, when they are viewed next to each other. On maintenance programmes, make sure that 'any old cheap tube' isn't used to replace old tubes in your sign.

Fluorescent tube replacement

In many cases, this fraud seems to be almost impossible to detect (although there is a partial solution). Furthermore, it is by no means restricted to the sign industry - in fact, it is highly probable that sign companies learnt the trick from disreputable building maintenance contractors.

Failed fluorescent tubes are one of the biggest clues to poor maintenance. Large illuminated signs – say, on the outside of an out-of-town superstore – may contain scores of these tubes; the signs look fantastic – bright, clean, clinical, with very precise brand representation. But the downside is that if one tube fails, the entire sign looks shabby, and since this is probably the most noticeable aspect of the store frontage, that's a problem.

Most retailers know that sites looking shabby, ill-kempt, and poorly lit, won't generate the same level of revenue as sites that look clean and bright. It's true of stores, petrol forecourts, motor dealerships and any other retail environment. So ensuring that the tubes are all working is critical.

It is almost impossible to buy fluorescent tubes with a warranty covering their expected life. But facilities managers know roughly how long tubes from their regular supplier will last in their particular application, and given their particular usage. The usual thing is for the sign maintenance contract to specify tube replacement, during normal maintenance visits, before the likely failure of the tube. In other words, if the sign is cleaned, inside and out, every six months, the visit immediately before the likely end of the tube's life will include tube replacement as a task.

But, unless you are there, personally, to oversee this, how do you *know* that the tubes are actually replaced? Even if you have very diligent, well-briefed, on site staff who try to monitor this, and you are confident that the tubes have been replaced, how do you know that the new ones have not been taken out of a building that the contractor worked on the day before?

CHAPTER SIX – DIRTY TRICKS

This is a scam which *may* be perpetrated by the sign company, as a matter of policy; or perhaps, both *your* company and the *sign* company are victims of the sign engineers, who are running a little illegal private enterprise of their own.

If this doesn't sound like a very profitable fraud to you, let me give you some 'back of a fag packet' figures. Let's say that a retailer has 200 sites. Each site has four big signs, containing 40 tubes apiece. That's 160 tubes per site, and 32,000 tubes across the retailer's estate. If it costs £5 a time (including the cost of the tube itself), 32,000 multiplied by £5 equals £160,000 saving. Put another way, they are charging you £160,000 for nothing. This is a fairly realistic scenario, and it does happen. Not a bad return for a bit of sleight of hand.

There is a partial solution, and one I have proposed for years. Adopt a colour for every year.

Every crew is provided with some pots of the correct paint, and a few small paint brushes. Every time a tube is replaced, a dab of paint (of that year's colour) goes on to either end of the tube. Subsequent spot checks, conducted randomly, are also built-in to the contract, and everyone knows this. The spot checks will reveal the year in which the tubes were replaced. At the end of the year, the colour is changed, for a different colour. You have to be strict about this: tubes which have no paint dabs have to be regarded as "failure to replace".

This simple technique will at least prevent them replacing tubes with old ones from your own properties, but it doesn't remove the need for you to be on your guard – there's no foolproof prevention method, that I am aware of, apart from diligence.

The directors of companies that carry out this fraud as a matter of policy must (presumably) believe that they are almost immune to prosecution. In order to conduct the fraud, they must tell employees not to replace tubes, and most sign engineers aren't daft – they'll realise what is going on. In normal circumstances, once knowledge of a fraud becomes widely known, the chances of detection increase rapidly.

Yet, to my knowledge, no sign company directors have been charged with fraud on the basis of this crime. I am told by friends in the police that many corporate frauds are never reported, because the victims prefer not to be publicly identified as having been taken in, so perhaps those that are found out are quietly sacked from the contract. Or perhaps it is an indication of the general standard of ethics in the industry – no one is surprised or shocked, because everyone knows how widespread the practice is.

Installation crews

Self-employed crews are often paid on a 'fixed price per job' basis, which means the risk of the job is passed on to the installer. Many installers are small companies, perhaps with only a handful of employees and with massive temptations to cut corners to save costs.

Routinely (and I use that word advisedly), such companies will avoid the use of access equipment, to save time and money. Installers, under pressure, may use unsafe practices, putting the customer (that is you) in peril of civil or criminal legal proceedings. Not all crews do this – but it is commonplace.

Component faults

Professional installation crews will report a fault on a component as soon as it is noticed. This will save you a fortune, because faults will be rectified *before* components have been fitted at all other sites, and return visits, months down the line become necessary.

Regretfully, the other solution, by less ethical companies, is to simply hammer the component into place, particularly if this repair is not visible to the customer. Why? It will, often, be done by freelance sign installation crews, subcontracted to the sign company, or by the sign company's own in-house crews, because no one gets paid until the job is completed – and reporting faulty components will lead to delay.

Structural support

Get a guarantee of the cement content used as the foundation in any large structural signs, such as those at the entrance to large retail sites, or MID signs on forecourts.

Using a low cement content is a sure-fire way to save the sign company money, but jeopardises the sign's structural integrity. Also, check that the teams use vibration rods, so that the concrete is properly compacted.

Check that your base workers have a clear understanding of the layout and location of manholes and ducting cable runs. Cable runs and ducting are different colours for safety reasons – if someone digs in the wrong place, you could lose *all* services, and someone's life could be endangered. Make sure they know this and that they use CAT (Computed Axial Tomography) scanners when necessary.

They need to have a solid grasp of the technical details relating to sign bases. For example, thread protrusion on bolts holding such a weight is critical (depending on the detailed circumstances, it should be a minimum of 1-3 full threads clearly protruding through the top of the nut). The nuts – usually eight of them – come ready mounted in a cage - it can be called a 'foundation cage', 'root cage' or sometimes a 'grillage', and its function is to ensure that the bolts are set into the ground at precisely the right distances from each other. The cage is literally sunk into the concrete.

Sometimes, however, the construction contractor will set the cage too deep, which means that the correct thread protrusion is not possible. There are two ways of dealing with this: rectify it (meaning delays to the contract, and delays to the sign company in getting paid), or plough ahead anyway, and decide not to worry about the lack of protruding threads. Installation firms tight on cash, and desperate to get paid, are under huge pressure to do the latter.

Electrical

Verification and certification is critical when it comes to electricians. With all electrical work, contractors should be trained to the highest relevant standards, and these should be checked. It is an area in which subcontractors can pull the wool over the main contractors (who don't always ask all the penetrating questions they should), and in which main contractors can also deceive the customer. Insist on

CHAPTER SIX – DIRTY TRICKS

seeing paper qualifications; you would be amazed how many times sign companies themselves don't insist on this. They haven't done a proper qualifications check, and when the customer asks for one, they can't satisfy the requirement. You should never accept 'improvers' on your sites, unless you know about them, the cost is adjusted accordingly, and they are being properly supervised.

Experience and credentials

Regretfully, some sign companies claim false credentials. Naming major brand customers as previous clients, when all they did was to fix the "gents" and "ladies" signs in the toilets in the Ashby-De-La-Zouche branch office, is misleading, and probably has little relevance to the job you need doing.

Make sure they have relevant, recent, and credible work experience, before you employ them.

Disreputable companies will also put major brands on their customer list even if they have not worked for that company for years. We came across a company which claimed a major petrol retailer as their customer, but the brand image they had worked on was nearly 12 years out of date!

Copyright abuse and theft

Make sure that the tender or presentation work submitted by a sign company has not simply been stolen from another – more reputable –

firm. You should pay particular attention to this if it is a complex or highly technical proposal. Some companies will acquire a rival's quotations, by devious means (possibly with the connivance of friends in your own company), and simply insert their own figures (yes, it *has* happened to my company). They may not even bother to change the wording, or the photographs and illustrations used in the original, which is a clear breach of copyright.

But shabby behaviour of this sort is fairly typical of the sign industry; the firm concerned may have had no *hope* of completing the job to the detailed specifications laid down in the stolen pitch document, but how would you, as the customer, know that, until it is too late?

Don't be surprised if ethical companies submit work to you with their copyright prominently displayed, and a declaration that the work is registered with www.protectmywork.com.

If you recognise, or suspect, theft of this sort, because rival companies are supplying identical words and pictures, get to the bottom of it. If a company is so unworldly that it believes it can steal words and pictures from a competitor with impunity, is it really fit to be trusted with your sign project? After all, if they are prepared to steal from a rival, and have crossed that moral boundary, what's to stop them stealing from you? If the company operates within a culture of ethics and integrity, such behaviour is unlikely to occur.

Faced with evidence of identical words and pictures, but different prices, the other possibility is an illegal cartel or ring, carving up business between its members and taking "Buggin's turn" to submit the lowest quote. In larger concerns, they may even have someone "on the take" inside your company. This *does* happen in the sign industry!

Copyright abuse and theft is a growing legal issue (ask your PR or advertising agency – at a senior level, they will know *all* about it, in great detail!) If one of your suppliers is stealing someone else's work (that includes pictures, illustrations, words, and anything that is their original creative work), and claiming it as their own, they are breaking the law, and / or risking a civil action. It's that simple.

Photographic deception

There is no reason why sign engineers should not take digital photographs of each completed site, as it is done, and post them on a website, where you can see them, instantly.

However, make sure that they take a variety of shots, from different angles. It has been known, for example, for pictures to be taken of a sign, and then used to justify works on more than one site. If you were managing a few hundred sites, would you notice? An insistence on a variety of views helps to frustrate fraud of this type. The biggest dirty trick is taking photographs of someone *else's* work!

Access equipment

Does the sign company always choose the appropriate access equipment for the job, or do they use trestles, ladders and rooftops, whatever is there, to save money? Check that equipment specified in the quote is actually provided. Some of the cowboys assume that you are just not going to be at every site, so how would you know?

Materials

De-speccing through choice of materials is a classic cheat. The manufacturers of, for example, vinyl graphics, will make a large range of apparently identical materials. Each specific material in a manufacturer's range is designed for a different application.

Durable, long term, outdoor graphics, will be more expensive – but if you are having a graphic sign, and you want it to last for long time, in all weathers, this is what you need. Make sure that the materials you pay for are right for the job.

Our advice is that if you have found a reliable sign manufacturer, who can provide sufficient answers to the most relevant questions given in this book, let them decide which materials you are going to need; they will also buy in bulk so that you get a better price. If you want to be better informed, visit the websites of the manufacturers of sign materials (such as 3M at www.3m.com, for example).

The factory

Firstly, does it exist? There are some people who do not produce signs themselves, but who have experience and contacts in the sign industry and will 'broker' your sign project. There is nothing wrong with this. It can take much of the hassle out of the project for you, and save you time. It isn't the way *I* would work, in this industry, but for some it may be the best option.

However, there are some brokers who kid their customers into thinking they are manufacturers. There is most *definitely* something wrong with this, not least, its downright dishonesty. Yet there are people who have got away with this for years (I know of one major UK company which bought signs for 20 years from a foreign supplier, without realising that it was a 'virtual' sign company and had no manufacturing capability of its own).

Brian Healy, of fraud investigators Haymarket Management Services, tells of one case where the customer thought he was paying for individually-made signs, at a cost which reflected that, and was eventually told (by Haymarket) that the supplier was buying them from a stock company, at a commodity price, and applying a huge mark-up before passing them on. I know of similar cases in the past. The potential gains can easily run into hundreds of thousands of pounds.

If all you want are standard health and safety warning signs, direction signs and other common formats, visit www.stocksigns.co.uk, or other, similar suppliers, and save yourself a lot of money. You only need to go to a sign manufacturer for custom-made signage carrying your company / brand name and identity, or an image which is individual and to your specific design.

On a significant project, you should *visit* the factory. Apart from verifying that it exists, you can get an idea of how neat and tidy the place is, and whether it looks well managed, or whether there are clear health and safety breaches all over the place. If they are prepared to breach health and safety regulations in their own factory, why would they bother to adhere to them on your premises at the installation stage?

The sharp-practice brokers will have an invisible margin on the job and know how to squeeze the manufacturer to a price. The manufacturer may well de-spec the job to get in to the price bracket. The broker will probably be savvy enough to know this, and know precisely how it was done; he may even be told, by the manufacturer. It is doubtful that this information will be shared with the customer, however.

Honest brokerage, in which you know the broker's margins, and you know who is producing your signs, is fine; dishonest brokerage should be exposed, and pushed out of the industry.

Corrupting *your* company

The slack contracts, service level agreements, and controls in the sign industry can be infectious. It is not unknown for sign firms to offer bribes and other inducements to decision makers, or decision influencers.

Sometimes, however, the infectious corruption in the industry seems to seep into the most well managed customer firms, and even tempt otherwise well behaved managers to go for the main chance. Fraud investigators Haymarket Management Services know of several cases in which senior managers in customer companies have applied (not

CHAPTER SIX – DIRTY TRICKS

very subtle) pressure to sign companies to employ their relatives! No overt connection is ever drawn between the offer of employment and the award of a contract, of course, but the connection is, nevertheless, clear.

Senior customer-side managers have also been discovered by Haymarket "stacking the deck" – seeking quotes based on onerous Service Level Agreements and other issues which mysteriously do not apply to one, fortunate company. The fact that one company has the odds in its favour is never made clear to the offending manager's bosses / colleagues, of course.

Brian Healy, of Haymarket, also recalls the case of the customer manager who felt the need for a prolonged comfort break whenever his favourite sign firm came in to discuss a potential contract. He would leave his guests in his office, and it wouldn't take them long to find the rival bid documents on his desk. By the time he returned, the competitor quotes would be back in their folders, and in due course, the favoured firm would submit a bid with prices suitably adjusted.

You could draw the conclusion that any supping with sign firms should involve extremely long spoons. In reality, I doubt that the corruption of your own management team is likely, if you represent a reputable company and have good controls and procedures for these things, but if there is to be fraud in the award of contracts, there is a high probability – in my view – that it will occur in the sign sector. Brian Healy concurs, and says the nature of the contracts, the culture of the sign industry, the fact that many customers do not understand this specialist niche market, and the size of the potential rewards, all conspire to create a tempting environment for fraud. You have been warned!

Chapter seven

Be a good customer

In many industries, customers get the level of service that they deserve, broadly speaking. In the sign industry, in my view, they may get a far worse service than they deserve; nevertheless, some customer behaviour can exacerbate the problems, so here are a few guidance notes on how to minimise the risk.

Don't change the ground rules.

Low-skills sign companies will not maintain chapter-and-verse on the changes you have made since the contract began. They will be aware, in general terms, that you have "mucked them about", and that this has had cost implications. They may try to recoup that money, but they may have difficulty supporting any particular sum with hard evidence.

Experience will tell them that you will resist this, because most customers will have far better financial controls, and will, not unreasonably, require evidence that the extra cost has been incurred legitimately. They won't have the hard evidence, and they will know that this is a probable cause of conflict.

Actually, the reason that historically, they will have encountered conflict in this area, is that they will have poor record-keeping, a slack original contract (which did not cover the situation adequately), and low negotiating skills.

There is a strong temptation for them to try to avoid "trouble", as they would see it, by recouping their loss surreptitiously, either de-speccing the contract without your knowledge, or "loading" invoices.

By contrast, a professional and well-run sign company will have excellent, computerised record-keeping, and strong negotiating skills, and they will negotiate with you from a position of some strength (however, they are also likely to have warned you, at an early stage, that your planned deviation from the contract may lead to additional cost, so that it is less likely to be a 'nasty surprise' and a cause of friction). Their original contract is much more likely to cover the situation and leave a lot less wriggle room.

Some customer Facilities Managers take the view that in the real world, on a large roll-out, contract deviations due to the unexpected are almost inevitable, and will openly declare that a contingency budget is in place for precisely this reason. This is simply good practice, and professional prudence.

That contract...

A sophisticated customer, dealing with an unsophisticated sign firm, would serve everyone's interests by insisting on good record-keeping as part of the contract, and by explaining carefully that additional costs will be met if a) they are flagged up in advance, and prior approval to incur them is sought, b) they are supported by good records, and c) they meet with the terms of the contract. Have the contract checked by your company solicitor.

Maintain a good audit trail

Follow-up all verbal communications with an email. Establish an audit trail of communications with the sign company. This makes "I-said-you-said" disputes less likely, and easier to resolve. Some sign companies are managed by people whose strengths do not lie in a literary direction, and the replies you receive may lack clarity, and precision, or may lose their meaning entirely in excess brevity, but at least a reply proves that your communication was received; if there is serious doubt as to whether it has been understood, follow-up with further emails (don't be tempted to explain on the telephone, and leave it at that).

Pay when you say you will pay

Those who have spent a lifetime working for major corporations, with a generous salary and benefits, comfortable office and a fully-expensed car, will have little understanding of those who live and work in a colder climate.

A small sign company will try to present an image of rock-solid financial stability, because no one wants to do business with 'losers'. The representative may turn up in a quality car (a big expenditure for a small firm, but one that is seen as a means of presenting an image). Behind the scenes, however, things may be a little less rosy.

The directors may have their houses on the line; they may have realised that they have taken on more than they can cope with; the intense pressure which situations like this cause can lead to marital problems and other family difficulties, redundancies and job insecurity, and health risks, for both them and their employees. The calm and joky character in front of you may be developing an ulcer – really – worrying about your company's late payment of invoices.

If you work for a large company, and you know full well that the suppliers' payment terms are regarded as a work of fiction, for the entertainment of your Accounts Department, then warn the supplier.

The more sophisticated and well-managed sign companies will be financially stable (but remember, you can check this), will have strong financial controls themselves, and will chase you for money, as they are perfectly entitled to do; but many sign firms will take a cold bath because of late payment on a large contract. They will pay additional interest on the money you *haven't* paid them (because they will be operating on borrowed cash), and they will have had to pay *their* suppliers for materials, before *your* company pays *them*.

Suppliers to the sign industry aren't daft – they know what the industry is like, and will often be very cagey about giving credit to sign firms, which can mean cash up front. It's a 'double whammy' for an already hard-pressed sign company.

CHAPTER SEVEN – BE A GOOD CUSTOMER

Privately, a well-directed sign company will be much more realistic; they live in the real world, may not even *expect* to be paid on time, and will have the financial muscle to withstand that, up to a point (they won't like it any more than anyone else).

It is *because* their financial controls are strong that they can afford to give you a better, and more reliable service – and it is for that reason that they will still be in existence years hence, perhaps to honour warranty claims, or to provide additional signage to consistent standards.

Poorly-managed firms may well expect you to pay late, but will be taken by surprise by the *extent* of the lateness; they will also exacerbate the problem by getting into contract disputes, because of their incompetence and/or your dissatisfaction, which will delay the payment further.

Have low expectations

This does not mean "settle for less"; quite the contrary. It means that if you are clearly dealing with a sign company suffering from the problems outlined in this book, then either change to another sign firm, or accept that (in order to avoid problems on both sides) you are going to have to educate them, and do some thinking for them.

Small sign companies can feel intimidated and frightened by large customers. Imagine being a salesperson, working for a small firm based on a scruffy industrial estate in a run-down part of the UK, and visiting the Facilities Manager of a global company in a glass and marble emporium in the middle of The City of London.

Some of the problems identified in this book probably stem from the resulting inferiority complex. Understanding this may help the relationship. They will often welcome a gentle nudge in the right direction, strong hints that you wish to have a *genuinely* good service from them, and that, on your side, the contract will be ethically managed; that you will pay, fairly, for a fair standard of work.

Spending some time gently leading them to the place you would like them to be is time well spent, and will be of mutual benefit. By having low expectations of the sign company, and managing your relationship with them accordingly (paying greater attention to detail, and the contract, than you might with any other supplier), you may end up with a more harmonious relationship, and a better outcome.

You probably don't need to do this with the market leaders, managed by ex-big company people who are at the head of substantial firms, but you might need to do it with a more typical sign firm.

Don't encourage a price-led mentality

If you want a quality job at a fair price, say so. Not once (because everyone says that, whether or not it really reflects their requirement); but many, many times. Say it until they believe it.

Explain to them that you will make every endeavour to check the details of their work, and that you want a quality job.

CHAPTER SEVEN – BE A GOOD CUSTOMER

Consider using a separate installation and maintenance firm

Working for some sign companies can be a little like being in the Army – long periods of boredom, followed by short periods of frenetic activity.

This is because only a relatively small number of major, nationwide sign contracts are awarded every year, and when a company wins one it needs to be capable of running the contract, and, perhaps, starting very quickly; yet, clearly, it is uneconomical to have large numbers of people and resources sitting around "in case" such a contract comes in.

The installation side is particularly challenging, because it involves employees travelling to remote locations across the UK. Most sign firms overcome this by employing large numbers of local freelance installers wherever, and whenever, they are needed. There's nothing wrong with using freelancers, but sub-contracting on this level poses multiple problems, not the least of them being the difficulty of achieving consistent quality, and stringent health and safety standards. Does the firm you are dealing with really have the skills to manage this?

A smarter move is often for the customer to employ an independent, nationwide installation and maintenance firm, and to do so directly (not through the sign manufacturer – leading installation firms may be reluctant to work for some sign manufacturers, anyway, because they will have bad experience of being paid; they will not want to be pressured to do a cheap job, quickly; and because working as a sub-contractor to companies with poor project management skills is a costly nightmare).

By employing the installation company directly, you have a better chance of getting objective feedback on things like 'fit' problems and other manufacturing-based difficulties; you will reduce the necessity for mark-ups; and you will be establishing a relationship with the kind of company that you will probably need, to maintain your signs over the longer term.

In some cases, you may find that the installation firm can also source the signs itself, and take responsibility for all project management; they may even do so with cost transparency, so the mark-up is transparent, and negotiated beforehand.

(I should declare a vested interest at this point: I run just such a company).

Prototyping

Many signs are not off-the-shelf, but bespoke products which have never been made to that particular design, or for use in that particular application, before. If this is the case, a professional sign company may suggest prototyping. This means testing new material and designs, in real-life situations, to ensure that they perform as expected. Materials manufacturers will have conducted field tests anyway – the results will often be available in product brochures, web sites, etc – but a sign is a combination of materials, and some may react against others. In some situations, you won't know, until you have tried. Even the degree of fading due to ultraviolet light may be an issue.

On a sophisticated sign programme, the sign company that suggests prototyping isn't trying to squeeze an extra few bucks out of you, it is being thoroughly professional. It is no good taking the view that "they are a sign company, they should know without needing to prototype." Quite often, there is no way *any*one could know. But make sure that they are experienced at prototyping, and that the programme follows good scientific practice, and is entirely visible to you. Better that you – and the sign company – discover a problem on a one-off prototype, than discovering it once you've installed several hundred signs across the country.

Guarantees, warranties, and liabilities

Guarantees and warranties are no good unless they have small print. Documents without small print may seem attractively simple and straightforward – "if it doesn't work, we replace it, period". In fact, the legal reality is that they are worthless – "ah yes, but obviously, we didn't mean that we'd replace it if you *abuse* the product!"

The manufacturer of the sign *materials* used should provide a guarantee or warranty on the materials (most of the larger, more credible materials manufacturers have quite good warranties, nowadays); and the sign manufacturer should provide a guarantee, or warranty, on the final assembled signs. They should both have lots of small print, relevant to the signs you have bought. They should cover resistance to weather (not just wind, rain, ice and snow, but sun – the major "killer" of some sign materials); they should cover a specified time period, the specific usage to which they will be put, and the maintenance programme.

We've already covered the contract, but beware that multiple contracts with many suppliers on a large project may allow room for problems to "fall between the slats". Suppliers falling out amongst themselves, and blaming each other, may create a maelstrom of claim and counter-claim which it is impossible to resolve, leaving you – the customer – as the fall guy. Having one prime contractor, who will take over-arching responsibility, is one solution; if the sign installer is also going to have the maintenance contract, and have an involvement with your sites for the foreseeable future, it makes sense to appoint them as the prime contractor. Any materials failure which occurs is then their responsibility; it may not be their *fault*, but it *is* up to them to sort it out with the materials manufacturer or the sign manufacturer. If you appoint a prime contractor, it is fair for them to make a percentage, to cover the risk transfer and associated costs, but there is no reason why you should not know what the percentage is. 10-15% is usually fair.

Know your stuff

It is probably a universal truth that informed buyers, buy better, but in the sign industry it is certainly true. Reading this book is a good start, for those new to the industry, but the more you know about the subject, the less likely you are to fall victim to shabby practices or sheer bad management.

CHAPTER SEVEN – BE A GOOD CUSTOMER

The industry does have trade magazines, and if you are spending any significant sums in the industry, reading them may well be a good investment of time.

Conclusion

Don't have nightmares

At the end of the popular BBC television show *Crimewatch*, the presenter, Nick Ross, frequently signs off with the words "don't have nightmares", and points out that the chances of being a victim of violent crime are small.

I would very much like to say something similar. Unfortunately, the chances of being a victim of the sign industry, at least to some extent, are probably fairly high. Having said that, if you act with caution, prudence, and professionalism, you may be able to limit your loss considerably, and you can write a small loss off as the cost of experience. If you find a well-managed, well-directed, *professional* and *ethical* sign company, and there are many of them out there, you may not even experience any loss at all, and I hope that this book has gone some way to equipping you for that task.

If the thought that it is difficult to secure quality work, for a fair price, leaves you with a pervading sense of injustice, then let me give you some other things to reflect upon. Many of those who practice the tricks and scams outlined in this book do not do so because they are evil individuals, and very often, they are as much losers as their customers.

Because it is relatively easy to set up in the sign industry, many people do so, and not all of them have the training, skills, mental preparation and other attributes necessary to be successful business leaders.

Frequently, they quickly find themselves out of their depth, in an overpopulated market; often, they have set up their businesses on borrowed money, and struggle to meet the interest payments. They have a vague understanding that cash-flow is important, but no one has taught them how to handle accurate financial projections; their ability to do this is massively impaired by the unexpected costs which arise from poor contracts, contract disputes, late payment, and so-on. This calls for higher-than-normal financial controls and skills, instead of which, many sign companies are generally deficient in this area.

The sign industry is also, often, deficient in other areas. Its environmental record is un-researched, but was almost certainly appalling in the days when screen printing dominated; it is probably only slightly improved in the era of digital sign making.

In theory, the quality of modern signage should be excellent. The showcase sign projects shown in the trade magazines are not made up, they are usually genuine; even an industry with endemic problems will have *some* triumphs, given sufficient critical mass.

However, as a general rule, the mantra 'let the buyer beware' is never more true than in the sign industry, and the book is an attempt to equip the buyer to be *very* aware.

Appendix 1

A quick checklist and matrix

Credentials

1. PASMA certificates for installers?
2. IOSH managing safety certificates (managers)?
3. IPAF (scissor lifts, powered access)?
4. ISO 9000?
5. The Blue Book (forecourts only)?
6. UK PIA certificates (forecourts only)?
7. Comp'Ex 7/8 (hazardous environments / forecourts).
8. CSCS skilled worker cards?
9. CDM regulations on-site?
10. Have any *managers* completed on-site training for crews?
11. Working at Height regulations, 2005?
12. IIF safety course?
13. NICEIC (for electrical work)?
14. City and Guilds 2380/2381 and 2391 (electricians / test and inspection engineers)?
15. IEE 16th edition Wiring Regulations (BS7671:2001 incorporating 2:2004 amendment) (electrical work)?
16. Clear procedures for engineers' versus electricians' remits.
17. EMC (electrical components and signs)?
18. International Contractor's Safety Risk System (what was your score?)
19. CIPS (purchasing managers)?
20. Cost transparency capability – proof?
21. Gainshare – proof?

22. Partnerships – proof?
23. Blue chip referees?
24. Awards?
25. Customer comments / feedback?
26. On-line access to job details?
27. Value engineering and design team?
28. Dedicated cost reduction team?
29. Reduced cost for customers – proof?
30. Large roll-out experience – proof?
31. Fast prototype projects – proof?
32. Ability to work with new materials – proof?

Financials

33. Credit history?
34. Profitable?
35. Sub contracted services as % of turnover?
36. Appointing and measuring sub-contractors – method?

Contracts

37. Confidentiality clause?
38. Confidentiality of sub-contractors?
39. Contract specific to this project?
40. KPIs and SLAs?

APPENDIX 1

Project management

41. Digital cameras for installation teams?
42. Good project management software and people?
43. Asset register of signware?
44. "Intelligent reporting"?
45. Measurement of success?
46. Continuous improvement – proof?
47. Supplier audits (quality of components / installations)?
48. "Making good" – what is expected?
49. Service delivery – show key performance data?
50. Project management availability?
51. More than one emergency number?
52. Crew vehicles – Tracker, SatNav/GPS?
53. How can you check on job progress at specific sites?

Health and safety

54. RIDDOR reportable accidents?
55. Monthly accident reports / record all incidents?
56. Risk Assessments, Safety method Statements, every job / site?
57. Training up to date?
58. Average paid training time per employee last year?
59. Employees rewarded for health and safety – how?
60. Specific health and safety targets?
61. Working at Height checklist?
62. Safe working environment – proof?
63. IIF?
64. Who audited Health and Safety Policy? Proof?

65. Does policy define responsibilities, organisational rules relating to risks?
66. Does policy detail alterations to leadership / administration / communications following promotions?
67. Does policy detail job planning and hiring?
68. Does policy detail employee management and training?
69. Does policy detail emergency preparedness, health control, PPE?
70. Does policy detail planned inspections, analysis, updates?
71. Regular reviews of Risk Assessments and Safety method Statements – how often?
72. Workers trained re: asbestos?
73. Are all operatives qualified first aiders?
74. Medical checks on employees before commencing work?
75. Minutes from health and safety committee meetings?
76. Minutes distributed / communicated to all employees? Proof?
77. Safety bulletins? Proof?
78. 'Near miss' culture?
79. Accidents, incidents and near misses all investigated?
80. On-site health and safety audits of teams?
81. 'Refusal to work' procedures?
82. Power tool audits / PAT?
83. British standards (components)?
84. Toolbox Talks – proof?
85. How are emergency procedures communicated?
86. COSHH statements?
87. Employees trained to spot hazards?
88. Induction training records for on-site personnel?
89. Disposal of hazardous waste – procedures, proof?
90. Job assessment for PPE?
91. Call-out facility 24/7? National coverage?

APPENDIX 1

92. Fire trace systems on vehicles?
93. Climate surveys?
94. PPE register?
95. Harness on all mobile elevated work platforms – proof?
96. Audits of safety harnesses?
97. Length of lanyards?
98. Harness register?
99. Risk control procedures?
100. Mobile scaffold tower register?

Appendix 2

Dirty tricks checklist

Fixings.

Fluorescent tubes – sufficient?

Fluorescent tubes – right type?

Fluorescent tube replacement – paint dab system?

Installation crews / price per job.

Component faults reported?

Structural support / construction skills?

Electrical credentials, verification, certification, trained for *your* work environment?

Experience and credentials – test them.

Photographic deception

Access equipment – genuinely being used?

Materials – correct spec?

Appendix 3

Useful website addresses

Giving web site addresses in a book is always dangerous. The sheer nature of the internet means that sites, or their addresses, may change. Hopefully, however, there is enough information here to find an appropriate website through a search engine, even if the address details are out of date by the time you read the book.

www.apea.org.uk
The Association for Petroleum and Explosives Administration (APEA) is a UK based organisation, drawing membership from all quarters of the petroleum industry, including Regulators from National and Local Government Authorities, Oil Companies, Equipment Manufacturers and Suppliers, Service and Installation Organisations, Training Establishments and many others. It is seen to be unique in representing all sides of the industry and in providing a forum for debate and the generation of technical guidance.

www.bsi-global.com
BSI Group is a leading business services provider to organisations worldwide, providing independent certification of management systems and products, product testing services, the development of private, national and international standards, performance management software solutions, and management systems training.

www.cenelec.org
CENELEC is the European Committee for Electrotechnical Standardization, a non-profit technical organization composed of the

National Electrotechnical Committees of 29 European countries. CENELEC's mission is to prepare voluntary electrotechnical standards that help develop the Single European Market/European Economic Area for electrical and electronic goods and services removing barriers to trade, creating new markets and cutting compliance costs.

www.cips.org
Website of the Chartered Institute of Purchasing & Supply. CIPS provides a programme of continuous improvement in professional standards and raises awareness of the contribution that purchasing and supply makes to corporate, national and international prosperity.

www.city-and-guilds.co.uk
The City and Guilds International website, which allows practitioners to access useful support materials including lesson plans, handouts and activities, revision help – even careers advice.

www.compex.org.uk
The website of Comp'Ex – the leading national training, assessment and certifications scheme for electrotechnical craftspersons working in potentially hazardous or explosive atmospheres.

www.coshh-essentials.org.uk
COSHH Essentials has been developed to help firms comply with the Control of Substances Hazardous to Health regulations.

www.cscs.uk.com
Website relating to the Construction Skills Certification Scheme. The organisation's aim is to help the construction industry get quality up, accidents down and cowboy builders out.

APPENDIX 3

www.energyinst.org.uk

The Energy Institute (EI) is the leading professional body for the energy industries, supporting almost 12,000 professionals both nationally and internationally.

www.environment-agency.gov.uk

Website of the leading public body for protecting and improving the environment in England and Wales. The Environment Agency aims to ensure that air, land and water are looked after by everyone in today's society.

www.haymarketco.com

A corporate investigation company with experience of fraud in the sign industry. They can also advise on fraud prevention and deterrence.

www.hse.gov.uk

The Health and Safety Commission is responsible for health and safety regulation in Great Britain. The Health and Safety Executive and local government are the enforcing authorities who work in support of the Commission.

www.ieee.org

Website of the world's leading professional association for the advancement of technology – the IEEE is the Institute of Electrical and Electronics Engineers. It is a leading authority on areas ranging from aerospace systems, computers and telecommunications to biomedical engineering, electric power and consumer electronics among others.

www.iosh.co.uk

Website of the Institute of Occupational Safety and Health – Europe's leading body for health and safety professionals with nearly 30,000 members worldwide.

www.ipaf.org

The International Powered Access Federation was formed in 1983 to represent the whole of the industry, from manufacturing, distribution and rental trades to associate businesses and users of all forms of powered access equipment.

www.iso.org

International Organisation for Standardisation – a network of the national standards institutes of 156 countries that ensures products or services conform to international standards.

www.niceic.org.uk

Information from the National Inspection Council for Electrical Installation Contracting – the industry's independent, non profit-making, voluntary regulatory body, covering the whole of the UK. The NICEIC's sole purpose is to protect consumers from unsafe and unsound electrical work.

www.opsi.gov.uk

The Office of Public Sector Information is at the heart of information policy, setting standards, providing a practical framework of best practice for opening up and encouraging the re-use of public sector information.

APPENDIX 3

www.pasma.co.uk

The Prefabricated Access Suppliers' and Manufacturers' Association website, with information on the safe use of alloy access towers and skills required. It devised the PASMA Training Course for users of alloy towers, now universally acknowledge as the industry standard.

www.pat-testing.info

Portable Appliance Testing is an important part of any health & safety policy. This site is intended as a guide to both the legal implications and to the technical requirements.

www.theirm.org

The Institute of Risk Management is the professional education body, established as a not-for-profit organisation, governed by practising risk professionals.

www.ukpia.com

The UK Petroleum Industry Association represents the oil refining and marketing industry in the UK.

www.3m.com

3M is a diversified technology company with a worldwide presence which manufactures the display and graphics materials used by sign companies.

www.stocksigns.co.uk

This site provides ready made signs for many applications. You only need a professional sign manufacturing company if you are having signs made to your specific brand / company design or image.

www.ingramcontent.com/pod-product-compliance
Ingram Content Group UK Ltd.
Pitfield, Milton Keynes, MK11 3LW, UK
UKHW021321180426
11947UKWH00015B/1363